Primates in the Real World

University of Virginia Press *Charlottesville & London*

PRIMATES
IN THE REAL WORLD

*Escaping Primate Folklore
and Creating Primate Science*

Georgina M. Montgomery

University of Virginia Press
© 2015 by the Rector and Visitors of the University of Virginia
All rights reserved
Printed in the United States of America on acid-free paper

First published 2015
ISBN 978-0-8139-3736-6 (cloth)
ISBN 978-0-8139-3740-3 (e-book)

9 8 7 6 5 4 3 2 1

*Library of Congress Cataloging-in-Publication Data
is available from the Library of Congress.*

In loving memory of my mother

Contents

Preface ix

Introduction 1

1 Separating Fact from Fiction 11

2 Venturing out of the Lab and into the Wild: Early Primate Field Studies 27

3 Control, Repetition, and Objectivity: Turning Field Observation into a Science 52

4 Capturing Natural Behavior: Changing Definitions of "Natural" in Mid-Twentieth-Century Primate Studies 71

5 Modern Primatology: The Emergence of Long-Term International Field Projects 91

6 Inclusion and Indigenous Researchers: The Africanization of the Amboseli Baboon Project 106

Conclusion 123

Notes 131

Index 155

Preface

This history book has an interesting history of its own. I was first exposed to primatology while a graduate student at the University of Minnesota. Sharing a house with two anthropologists, I quickly found myself socializing with various anthropologists on campus and became friends with people who worked at the Jane Goodall Institute Research Center (JGI). Alongside my history of science courses, I studied primatology and animal behavior in graduate classes taught by Anne Pusey, the director of the JGI. I developed friendships with people for whom it was second nature to see photos of dozens of Gombe chimps and individually recognize and name each one. I remember attending a gathering at Pusey's home with Jane Goodall and realizing this cultural icon was simply "Jane," a friend and colleague to many in my social circle.

My social connections with primatologists were further solidified when I met my partner, Bob, in Pusey's primatology course. Bob worked for several years as a student at the JGI and had recently returned from fieldwork at Gombe. Pusey still recalls that it was she—or at least her class—that brought Bob and me together. Although no longer focused on primates, my husband continues to work in wildlife biology and ecology. As this brief autobiographical account makes clear, *Primates in the Real World* was born in a lived reality, during a time when my academic interest in primates and primatology integrated with my social and personal life.

With one foot in the history of science and another in primatology, I was comfortable reading literature and talking with colleagues from both the humanities and the natural sciences. I was also aware that throughout the 1980s and 1990s the so-called science wars encouraged hostility between the humanities and sciences. In more recent years, many historians of science and scientists have sought to bridge these "two cultures" to foster conversation and collaboration across disciplines. Within primatology, however, the science wars had particularly caustic effects in part due to the strongly negative reaction many primatologists had to Donna Haraway's 1989 book *Primate Visions: Gender, Race, and Nature in the World of Modern Science.*

Primate Visions is widely recognized as a seminal work in the history of science, science studies, and gender studies. With chapters dedicated to leaders in the field, Haraway offers both chronological breadth and analytical depth with a focus on gender and social control issues. Unfortunately, many primatologists found it to be overly driven by a feminist agenda, resulting in what they perceived as a narrow account of their discipline.

Reviews of *Primate Visions* in primatology journals were usually strongly critical. Two years after its publication, Matt Cartmill from the Department of Biological Anthropology at Duke University, for example, reviewed *Primate Visions* for the *International Journal of Primatology*. Cartmill's book review is extraordinary. It includes statements such as the following: "This is a book that clatters around in a dark closet of irrelevancies for 450 pages before it bumps accidentally into its index." And simply: "This book infuriated me." Cartmill's review certainly is not a warm one, and it is written in response to what he calls Haraway's "unfriendly" approach, "which makes no effort to understand or to sympathize with the intentions of scientists" and indeed "caricatures" primatology "into unintelligibility." Cartmill does concede that "there are real insights and intermittent flashes of brilliance" in the book, and he says "all primatologists will benefit from reading it." Anthropologist Peter Rodman gave a warmer account when he reviewed the book for *Current Anthropology* in 1990; however, even he concludes that "it is unlikely that anyone described in *Primate Visions* would belittle Haraway's stakes—anticolonial, antiracist, antisexist, and antiwar—and we cannot blame a colleague for having stakes, especially when they are made so clear. At the same time, scholars cannot take this biased story of primatology seriously."

In many ways this episode from the late 1980s and early 1990s is emblematic of the science wars writ large. Historians of science, particularly those mistakenly labeled as antiscience because of their feminism as well as those interested in exploring contemporary science, may still encounter scars from the science wars. Scholars examining primate studies are no exception.

Chapter 6, which turns to more contemporary primatology, includes interviews of researchers I conducted while visiting the field camp of the Amboseli Baboon Project, a multidecade, ongoing primatology project in Kenya. This visit and the oral histories it generated would not have been possible without overcoming the distance that the science wars created between primatologists and historians of science. It took me five years of emails and meetings with Jeanne and Stuart Altmann, who founded the Amboseli Baboon Project, and an interview with Susan Alberts, current codirector of the project, at Duke University to build a relationship with them and be invited to the Amboseli Baboon Project field site. Visiting Amboseli was essential for completing this book, because no other history of primatology had included the voices of African researchers and I believed it was essential to correct this silence.

Before and after my visit to Amboseli, Susan Alberts served as my main point of contact with the Amboseli Baboon Project. My professional and personal connection with her began when I interviewed her at Duke University after I gave a talk for their Women's Studies colloquia. Alberts and I have developed a close professional relationship as a result of our time in the field together, which, although brief, was significant for us both. This was in part because I visited shortly after several of the baboons in their study were killed, which was an emotional time for the project's researchers.

Alberts—along with Raphael Mututua, Serah Sayialel, Kinyua Warutere, three Kenyan researchers who are permanent employees of the project, and Courtney Fitzpatrick, a graduate student who was also at Amboseli when I visited—welcomed me to the field camp and gave me a lot of their time. In preparation for interviews with the researchers, I had read methodological works from anthropology. Some of these books included reflections on the advantages and disadvantages of being a participant observer, including the inevitable emotional connections that form between observers and subjects.

I see many parallels between my own experience writing this book and the primate researchers I have examined to produce it. Like me, they form emotional connections with their study subjects that are essential for their work and thus should not be dismissed as merely subjective and thus inaccurate. Furthermore, as this preface has served to elucidate, this book grew from a merger of my professional and personal life. Although this no doubt resulted in elements of subjectivity in my analysis, I strongly believe that this book would not have been possible without my relationships with and respect for members of the primatological community. Certainly, without such relationships I would not have been able to incorporate Mututua's, Sayialel's, and Warutere's perspectives into my narrative. I also would not have had the unusual opportunity of sitting among a group of wild baboons. Having read about primates' ability to communicate warnings, I actually saw baboons standing to warn each other of a nearby snake and in turn benefited from this warning myself. Visiting Amboseli also allowed me to meet and then name "Eugene," a new male that joined one of the study groups while I was there.

I hope that historians of science and primatologists alike will find this book engaging and valuable and will find added value in my use of personal experience. The book will surely differ from histories of primatology that focus on controversies in the field and the intellectual context in which theories of primate behavior developed. Perhaps in part due to my connections with members of the primatological community but more so due to a desire to avoid repeating what has already been done by others, I choose to avoid revisiting well-covered areas. Instead I explore the history of primatology through a series of episodes focused on the people, practices, and popular culture that shaped the development of this remarkable scientific field to help demonstrate how primatologists professionalized their field and how today the field is being reshaped by new goals and new pressures.

Many colleagues, friends, and family members helped me on the journey involved in researching and writing this book. I am grateful to Paolo Palladino for introducing me to the history of science. While at the University of Minnesota I benefited greatly from the guidance of John Beatty, Sally Gregory Kohlstedt, Gregg Mitman, and Anne Pusey. During the years while I was writing and revising, several other fellow historians of science also kindly read chapters or the whole manuscript including Harriet Ritvo, Erika Milam, Audra Wolfe, Richard Burkhardt Jr., John Waller, David Bailey, and Helen Veit. Many primatologists also generously spent time talking with me and reading my work, including Susan Alberts, Stuart Altmann, Jeanne Altmann, Anne Pusey, Mike Wilson, and Ian Gilby. During my visit to Amboseli, Kenya, where I interviewed members of the Amboseli Baboon Project, Courtney Fitzpatrick, Raphael Mututua, Serah Sayialel, and Kinyua Warutere welcomed me with open arms. I will always be grateful for them allowing me to experience their field camp.

While researching this book, I benefited from the hard work of archivists at Yale University Library, Pennsylvania State University Archives, San Diego Historical Society, archives at the University of Arizona, and the British Library. Parts of this book appeared in previous articles of mine, and I thank the *Journal of the History of Biology* and *Endeavour* for permission to include them. I am also incredibly grateful for the guidance and support of Boyd Zenner, Mark Mones, Angie Hogan, Morgan Myers, and Michele Callaghan at the University of Virginia Press and for the two anonymous reviewers who read my manuscript.

Several friends and colleagues provided encouragement at important moments, including Rachel Mason Dentinger, Gina Rumore, and Cori Fata-Hartley. Mark and Brie Largent provided me with incredible friendship and support throughout this journey, and I will always be grateful to them both. In particular, Mark Largent read the entire manuscript more than once and the book is much improved as a result of his comments.

Outside of academia, my lifelong friends Amy Sharpe, Rachel O'Flaherty, and Nicola Thurston supported me in this endeavor as they have in everything. I also want to thank my dad, Colin Hoptroff, his wife Ann, my brothers Ali and Mike, and Lee and Sheila Montgomery for their love and support. Although my mother died around the time this project began, I owe her an incredible debt. Her unconditional love and continued belief in me continues to inspire me. Finally, I would like to thank my partner, Bobby, who shared every high and every low from the start to the finish of this project. More than anything I thank him and our incredible daughter Olivia for filling my life with laughter and adventure.

Primates in the Real World

Figure 1. Photo of baboon graves in the Amboseli field camp. (Photo taken by author)

Introduction

The graves, each crudely marked with a stick for a headstone, greeted me when I arrived at the Amboseli Baboon Project in Kenya in 2009 (see figure 1). Two months earlier, amid one of the worst droughts ever to hit the region, local Maasai had speared seventeen baboons that were part of the Amboseli Baboon Project's study groups. The unprecedented act of violence was retaliation after baboons killed several goats, which are valuable economic and social commodities in Maasai culture. Kenyan researchers, Raphael Mututua, Serah Sayialel, and Kinyua Warutere, found eleven of the baboons' bodies and buried them within the grounds of the field camp.

When they talked to me about the killings, the Kenyan researchers expressed great love for the primates they studied, identifying them as "very much like us" and indeed "part of us." For Mututua, Sayialel, and Warutere, the primates' human-like qualities fostered a deep emotional connection and warranted the baboons' burial as a mark of respect.[1] Not only did Sayialel find, collect, and bury the bodies of the baboons, she also observed the impact of the killings on the baboons' family: "We felt very bad because you know it [the baboon group] is like part of us . . . but we really felt bad because when they killed them in March . . . they really hit a family which was really high ranking. It was one of the D families, which was one of the high-ranking families for quite a long time and it affected almost the whole family."[2]

Susan Alberts, one of two American primatologists who direct the Amboseli Baboon Project, provided an explanation of the burial that was simultaneously very different and very similar to that offered by the Kenyan field team. It was, she said, a scientific study of bone decomposition, rather than an act of memorialization of the primate participants. However, this seemingly objective, scientific act had an additional motivation. She explained that the researchers had decided to conduct the study "to make the baboons' deaths mean something."

The act of burial is a ritual that for humans encompasses respect, remembrance, and a sense of closure. It embodied the baboons' roles as scientific subjects and unique individuals toward whom these researchers felt emotional attachments and ethical responsibility. Alberts described her profound sadness at the loss of the baboons within this framework of responsibility: "I feel that I failed the baboons. . . . Really one of the only things that I potentially have to offer the baboons in exchange for the innumerable and inexpressible ways they have enriched my life is protection from this kind of killing, and I failed utterly to do that."[3]

The baboons' deaths and burials—as well the emotional and scientific responses of Alberts, Mututua, Sayialel, and Warutere—illustrate the legacy generated by the long and complex history of primate field studies. They allow us to explore the interaction of values of scientific objectivity and distance with a lived reality involving long-term studies of the lives of individual animals that are in many ways much like us. Transferring the scientific value of detachment is particularly problematic in modern field primatology, because contemporary projects require prolonged and often intimate relationships with individual animals and their families. However, this combination of science and affection is not entirely new. Indeed, some of the earliest studies of captive primates relied on a strong bond between primates and researchers. The epistemological tension created by valuing distance while requiring intimacy helps explain several aspects of the researchers' responses to the baboons' deaths, but it is only part of the story.

The Birth of a New Science

Unpacking the collision of the extreme drought, local people, Kenyan researchers, and Western scientists engaged in primate studies at the time of the animals' deaths reveals many of the values, actors, and places that directed the development of modern field primatology.[4] The history of primatology has long challenged conceptions of what it means to be a scientist and a science. In the twentieth century, American primate researchers had to be mindful of distinguishing themselves—their identity, methods, sites, and ideas—from the amateurs and the travelers' tales that came before them. In the pursuit of scientific credibility, field scientists increasingly embraced experimentalism while perpetuating a particular concept of naturalness as the defining characteristic of their work. In time, the tension created by manipulating both primates and their environments, while simultaneously claiming naturalness, inspired a group of primatologists to pursue noninterventionist field studies in the primates' native habitats. To do so, modern primatologists like Alberts had to take on the methodological, logistical, and political challenges created by doing long-term, transnational fieldwork.

Many aspects of the history of field primatology revolve around distance. In particular, early primate researchers sought to create distance between the popular primate stories written by amateurs and adventurers and the more orthodox primate science performed by scientists. At the same time, however, creating a science based on studying primates necessitated working with people and in places that fell outside of traditional definitions of "good science" and with animals that in many ways appear human. Primatologists both past and present had to maintain, and in time expand, their relationships with individuals outside the traditional scientific community. Thus, in many ways primate studies remained enmeshed with the work of so-called amateurs and popular representations of primates and primatology even after the field had attained the status of a professionalized science. Pursuing primatology also meant keeping the animal subjects close, both physically and emotionally. Primatologists keep their subjects close enough to observe them for prolonged periods of time and close enough to the heart to warrant memorializing them.

This book explains how the qualities that make primates so interesting and informative are the very characteristics that made the professionalization of field primatology so challenging and unusual. The standard history of science in the United States traces the replacement of amateurs with self-identified professionals during the nineteenth century.[5] Unlike many scientists in the United States, American primatologists had to wait until well into the twentieth century to establish themselves as scientific specialists, and even then they remained dependent on individuals outside the traditional scientific community for a range of information and practical necessities. A remarkably young field—with societies like the International Primatological Society and the American Society of Primatology being founded in 1980 and 1981, respectively—it took considerable time for primate researchers in the United States to create a distinct and legitimate epistemological space for themselves that could support the common hallmarks of scientific professionalization, which includes societies, journals, funding sources, and graduate programs.

The shadow cast by centuries of what I call *primate folklore*, a hybrid of tales, facts, and fictions about nonhuman primates and their physical and behavioral similarities to humans, represented the greatest challenge for professionalizing primatologists. The scientists who took on the task of replacing the primate fictions of amateurs with the primate facts of professional scientists in the early twentieth century recognized that they were countering an image of primates that had been etched into the collective consciousness of the Anglo-American public over hundreds of years. But they also had to change the public's conceptions about who could properly observe primates in the wild, while at the same time preserving a range of collaborations with individuals who lacked formal scientific training but nevertheless made essential contributions to studies of primate behavior and biology.

The first three chapters detail the widespread nature of primate myths and stories in the nineteenth and early twentieth centuries and explain how a small group of American primatologists in the interwar period worked very deliberately to supplant the many primate fictions that animated primate folklore with findings from the new science of primatology. Many of their efforts focused on establishing methodologies that would mark their work as scientific rather than as sensationalistic. Chapters 4, 5, and 6 explore how modern primatologists integrated new methods and responded to new demands as the field came into its own during the latter half of the twentieth century. During a time when field studies began extending to multiple decades, primatologists' research attracted a new form of primate folklore for the modern age, with magazines, films, books, and blogs creating what I call *pop primatology*. The limelight fell on a select group of researchers, but all field primatologists in the postwar period were faced with new research opportunities and new logistical challenges. With the growth of long-term field studies, many American primatologists began employing indigenous people as assistants and managers. For some, directing transnational field teams demanded new ways of expanding who was included and given due credit in primate studies. Combined, these methodological developments and the later efforts to expand participation resulted in a field science with diverse contributors working across continents to ensure continual observation of wild primate groups.

The Burden of Popularity and the Process of Professionalization

The popularity of primates may appear to hold certain advantages for primate researchers. After all, celebrity status, funding opportunities, employment possibilities, and easy dissemination of findings often go hand-in-hand with one another. However, in the years that primatology sought the hallmarks of professionalism, it had to confront its own complicated entanglements with popular culture. In the early twentieth century, the public's appreciation of all things primate, perpetuated by the illicit thrill of gazing upon beasts that were simultaneously familiar and exotic, created substantial challenges for the fledgling science of primatology.

The professionalization of scientific disciplines, as with any other professionalization process, necessarily involves creating distinctions between amateurs and professional scientists. The transformation of natural history into modern biology, for example, involved leveraging academic credentials, methods, and equipment to brand worthy naturalists as professional scientists. The emerging professional naturalists recognized amateurs as providers of materials and raw data but claimed special privileges over the creation of knowledge based on the analysis and evaluation of amateurs' data.[6] Separating professionals from amateurs was never easy, but it proved particularly difficult for the emerging science of primatology, because primate folklore had long emphasized the expert knowledge of informal observers and even systematic amateurs. Moreover, the rarity of certain primate specimens made emerging professional primatologists unusually dependent on local knowledge and well-traveled amateurs for finding, observing, and collecting primates in the wild.

The American psychologists and anthropologists first drawn to the scientific study of our primate cousins in the early twentieth century were an eclectic mix of mavericks who showed relatively little concern about disciplinary boundaries. Many of them were willing to create innovative collaborations, methods, and sites with which to build the science of primatology. Robert Mearns Yerkes, who captured his interdisciplinarity in the term *psychobiologist*, leveraged his professional connections and his position at Yale to ensure that early primate field studies received funding and institutional support. Yerkes initiated studies of natural primate behavior to supplement his captive observations, and, while he focused on learning how primates lived, reproduced, and behaved in captivity, he funded younger researchers to explore, and at times endure, the wild primates' jungle habitats. Clarence Ray Carpenter was one of these younger researchers. A comparative psychologist with a background in bird behavior, Carpenter was the third postdoctoral fellow employed by Yerkes to conduct field studies of primates and the first to conduct a scientific field study of primate behavior. Recognized in his own time as a trailblazer, Carpenter and individuals like Yerkes played instrumental roles in the development of the science we now call primatology.[7]

From the outset, provocative questions about intelligence, sex, social relationships, communication, and tool use formed the core of modern primatology. Intellectually, exploring these topics involved drawing on cognitive science, linguistics, sexology, and sociology. Practically, it demanded enduring the heat, illnesses, and mosquitoes that often characterized field research. Researchers required almost infinite patience to do sustained observations of primates in the wild. But this reality was also mitigated by the fact that researchers did not endure these conditions by themselves.

Despite the images frequently constructed in their scientific papers, popular articles, and private field notes, the male researchers who conducted the first scientific studies of primate behavior were far from alone in the wilderness. At times their wives accompanied them, providing food, making camp, and occasionally observing and photographing primates alongside their husbands. Other women drew from their experience with captive specimens, providing vital information about basic primate biology and reproduction as part of the male researcher's field training. The male Western researchers also consistently relied on indigenous peoples as guides, porters, and gun-bearers when they ventured into remote primate habitat. Although such practical contributions were publically recognized and valued, scientists depicted indigenous peoples' observations of primates as unreliable, which freed American naturalists to claim for themselves the first sightings of the exotic animals that populated the newly colonized lands.[8] This was as true for the nineteenth-century stories of hunter-adventurers as it was for the primatologists who conducted field expeditions in Africa and Asia in the 1920s and 1930s. Indeed, such separation of local knowledge from expert observation continued to characterize field primatology throughout most of the twentieth century.

The complex infrastructure of Western scientists, indigenous assistants, trackers, porters, wives, and female collaborators formed a dynamic web of expertise. As an emerging field science that required collaborations among diverse individuals, identifying a particular individual's role as teacher, apprentice, assistant, and servant often was not easy, making the assignment of credit messy at best. Historians of science like Bruno Strasser and Robert Kohler have used the term *moral economy* to examine how status, credit, and authorship were distributed among scientific colleagues in the laboratory.[9] In primate field studies, these forms of power and recognition intersected with issues of race, gender, and colonialism, while also hitting upon distinctions between professional and amateur.

All too often, the professionalization process involved separating out those with newly certified expertise from others deemed inferior, a status that frequently correlated to individuals' sex, race, class, or nationality. For example, during the transformation of midwifery to gynecology, male doctors replaced female midwives as practitioners.[10] This traditional route to professionalization was not an option for primatology for a number of practical reasons. In the early years, amateur women provided male scientists with access to rare captive specimens. Later, with the appearance of Jane Goodall, Dian Fossey, and Biruté Galdikas—who were known as "Leakey's Ladies"—women were considered especially well qualified for observing primate behavior. In modern primatology, women emerged as powerful professionals, running multidecade field projects, developing new methodologies, and providing an example for what other sciences can do to attract, retain, and promote talented female scientists.

A gradual movement to credit indigenous peoples' contributions to primatology has followed several steps behind the increasing role of women in the field. Although local peoples' part in early primate expeditions was typically evidenced by occasional references to them as porters or guides in scientific publications or archival photos of unidentified men carrying equipment or holding captive primate specimens, starting in the 1980s primatologists sought ways to increase the visibility and recognition of indigenous peoples' contributions to primate research. Today, local people frequently work as researchers, making prolonged observations of primate behavior or as field managers who combine observational responsibilities with negotiating various logistical challenges involved in long-term fieldwork. Such contributions are increasingly recognized in the form of co-authorship on papers and more visibility for indigenous researchers on project websites.

The principal subjects in the development of field primatology have, of course, been the primates themselves. Their personalities and habitats determined which species were readily observable in the wild, which had to be studied in captivity, and what observational practices were even possible. Thus, primates were in many ways essential contributors to the development of primatology, albeit as unconsenting subjects of the new science. In primatology as practiced by Western scientists, primates' participation has been masked by a historic aversion to anthropomorphism, without which it is difficult to recognize primates as essential contributors to the science about them. Japanese primatology, in contrast, has always embraced anthropomorphism and has openly recognized that animal subjects are true participants in primatological research. Indeed, when a study subject dies, Japanese researchers perform a memorial service to thank the primate for all he or she gave to the project.[11]

And, as essential parts of any primate study, the animal subjects exerted agency in a whole host of ways. The rarity of certain primates, the difficulties involved in keeping them alive in captivity, and the challenges involved in seeing them in the wild forced scientists in the 1920s and 1930s like Robert Yerkes to be open-minded about their collaborators and about what constituted appropriate sites for science. Before the 1950s, researchers who wanted to study gorillas had little option but to turn to captive specimens. Those who wanted to conduct wild studies of howler monkeys, for example, had to be willing to travel to their habitats in South America, and any study of gibbons' natural behavior demanded expeditions to Asia. In this respect, the primates themselves determined where primate studies were conducted. In a similar fashion, primates' temperament shaped the methodological development of primatology. The curiosity of gorillas allowed researchers to finally overcome the animals' shyness and coax them out of hiding. Similarly, the wariness of many primate species demanded that scientists invest weeks, if not months, habituating the animals to human observation.

Establishing field primatology as a professional science meant negotiating a number of unanticipated hurdles, including the participation of sometimes noncompliant primates, the continual need for collaborations with amateurs, and the often central roles played by women and indigenous peoples. But challenges also brought opportunities. The primate participants attracted public attention, which in turn led to funding from a variety of nonprofit groups and foundations, including the Rockefeller Foundation. Collaborations with hunter-collectors and with women working at private collections and zoological gardens provided early primate researchers with invaluable access to animals at a time when it was understood to be difficult, even impossible, to observe them in the wild. Relationships with indigenous peoples were essential for the logistical operations of early primate field expeditions, and these relationships would later evolve, as we shall see, into the lifeblood of long-term modern field primatology projects by making possible the day in and day out observations that frequently spanned both decades and continents.

The need to form a contrast between field primatology and the stories and myths that had come before shaped how women, indigenous researchers, and the primate subjects were integrated into American field primatology. It was not easy for researchers to escape the cultural constructions of how primates supposedly looked and behaved. Never had it been more important for an emerging science to demonstrate its objectivity and accuracy, to declare to scientists and the American public alike that this was real science, wholly distinct from the adventure stories that had come before it.

1

Separating Fact from Fiction

Primates have been represented in American popular culture, both historically and today, more than perhaps any other animal. Americans' curiosity about primates' appearance, capabilities, and temperament has stimulated a flood of legends, images, cartoons, and films, some of which were completely fabricated while others wove together facts and fictions. Popular narratives and caricatures fanned the flame of public interest in primates and created the momentum needed for primate stories to continue for centuries.

The many myths about primates meant that scientists wishing to study primate behavior had to work especially hard to distance their work from the ever-popular stories about primates. Thus, what may first appear as an opportunity to enjoy the luxury of constructing a science around a topic already publically valued was in reality a heavy burden that took time and tactical effort to overcome.

The misinformation and exaggeration propagated by people who purported to have traveled to exotic lands and to have seen primates in the wild prompted a small group of American scientists in the early twentieth century to seek accurate, scientific knowledge about primate biology and behavior. To put their work in context, it is important to first understand what the American public and their fellow scientists were seeing and reading about our primate kin and the people who observed them. Like the apes featured in these popular primate stories, these accounts created an "Other" from which early primate researchers sought to distinguish themselves and their research.

Primate Tales, Myths, and Stories

Primates are charismatic creatures that have long garnered the attention of zoologists, naturalists, adventurers, poets, writers, and even the general public. In 1699, English comparative anatomist Edward Tyson dedicated an entire text to the similarities and differences between man and the orangutan, which was most likely actually a chimpanzee.[1] He declared that in contrast to other apes, the orangutan brain was "abundantly larger . . . and in all its Parts exactly formed like the Humane Brain," its liver undivided, as in man, and its heart "not so pointed, as in Apes." Of course orangutans also differed from humans in a myriad of ways, including "the littleness of its Stature[,] . . . the flatness of the nose," and "in having no pendulous Scrotum."[2] The close connection between primates and humans was also reflected in Tyson's images (see figure 2), which showed orangutans in human-like poses that were reminiscent of Andreas Vesalius's anatomical drawings from his 1543 work *On the Fabric of the Human Body*.[3] By the late eighteenth and early nineteenth centuries, naturalists like Johann Friedrich Blumenbach and Charles White supplemented such vivid accounts of primate anatomy with descriptions of primate behavior in which male apes acted as sexual aggressors who attacked human females. White even went so far as to assert "that women have had offspring from such connection."[4]

While scientific observers explored the specific similarities and differences between humans and primates, literary authors projected them in roles ranging from criminals to symbols of humanity. When in need of a villain for the 1843 story "Murders in the Rue Morgue," the American writer Edgar Allan Poe chose an orangutan. Across the Atlantic, the Scottish playwright and novelist Sir Walter Scott selected an orangutan to play a central role in *Count Robert of Paris* (1831) and used it as a convenient surrogate for exploring which human vices were and were not part of our "natural" selves.[5] Other primate literary characters, like Sir Oran Haut-ton and Mr. Chimpanzee, breached the human-animal boundary, ascending to the role of politician in Thomas Love Peacock's 1817 *Melincourt* and cultured traveler in "The Disappointed Traveller," a satirical short story written under the pseudonym Alfred Crowquill in 1840 (see figure 3).[6]

The publication and discussion of Darwin's theory of evolution by natural selection took the steady interest in primates that had characterized the seventeenth, eighteenth, and early nineteenth centuries and amplified it into a cultural obsession. By the late nineteenth century, Darwin's *On the Origin*

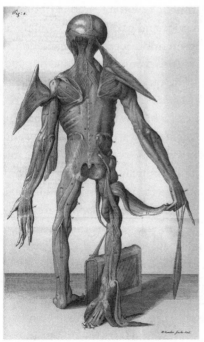

Figure 2. Drawings of an orang-outang from Edward Tyson's 1699 *Orang-outang, sive Homo Sylvestris: or, the Anatomy of a Pygmie.*

Figure 3. Mr. Chimpanzee from Alfred Crowquill, "Mr. Chimpanzee, the Disappointed Traveller," *Bentley's Miscellany* 8 (1840): 490–93.

of Species, Descent of Man and Selection in Relation to Sex, and *The Expression of Emotions in Man and Animals* were on scholar's bookshelves as well as middle-class coffee tables. At the same time, Darwin inspired cartoons in popular periodicals that were enjoyed in pubs and parlors. Unlike many of his nineteenth-century counterparts, Darwin fully accepted humans' continuity with other animals. His 1859 *Origin of Species* only implied that humans descended from lower order animals, but he tackled the topic of humans' evolutionary origins directly in *Descent of Man* in 1871 and even more forcefully in its sequel, *Expression of the Emotions,* which was published just one year later. Continuity among all animals lies at the heart of both *Descent* and *Expression,* with Darwin declaring any difference between humans and animals to be "one of degree not kind."[7] Name any human characteristic—sympathy, love, morality—and you will find some form of it in other animals, Darwin said. He expanded on this declaration in *Expression,* where every line and every image demands the reader accept his contention of the self-evident continuity between man and animals.[8]

For Darwin, mental experiences like love, dreaming, sadness and sulking were literally etched onto the body. A human or animal's face, body, or posture exposed the subject's mental state to the careful observer (see figure 4). It is telling of our discomfort about the consequences of the assertion that we are truly animals, and the creatures that we domesticate, hold captive, and consume are much like us, that *Expression* remains so rarely referenced and so relatively unknown.[9]

Throughout the latter half of the nineteenth century, primate stories fused with Darwinian evolution. Primates, particularly gorillas, attracted more attention than ever, taking the form of characters in literary works such as Richard Grant White's *The Fall of Man, or The Loves of the Gorillas, A Popular Scientific Lecture upon the Darwinian Theory of Development by Sexual Selection, by a Learned Gorilla* (1871) and motion pictures such as *King Kong* (1933).[10] In the wake of Darwin's work, these literary primates reflected

Figure 4. Three sketches from Charles Darwin, *Expression of Emotions in Man and Animals* (1872). From left to right: dog "caressing his master," "cat in an affectionate frame of mind," and "chimpanzee disappointed and sulky."

not only society's real anxiety about the human-animal boundary but also its beliefs about the naturalness of man's sexual aggression and the worryingly close connection between women and humans' animal origins. By the late nineteenth century, Darwinism had provided the necessary catalyst to propel the stories and speculations of primate biology and behavior into a widespread cultural phenomenon known as the "gorilla craze."[11]

Gorillas took center stage in texts and images that embodied Anglo-America's simultaneous anxiety and fascination about humans' primate kin throughout the late nineteenth century. Stories, travelers' tales, poems, cartoons, and illustrations of primates reflected Victorian concerns about the human-animal boundary, particularly as it informed issues related to gender and race. The poem "The Missing Link," published anonymously in the *Boston Commercial Bulletin* and *Ward's Natural Science Bulletin* in 1880, for example, chronicled a fictional wedding between a black woman and a gorilla that produced "Mr. Darwin's missing link."[12] Some individuals were labeled as missing links in an attempt to identify them as subhuman due to their race, gender, or nationality. Sara Baartman is perhaps the best-known victim of such marginalization and exploitation, suffering display as a "Hottentot Venus" in England and France between 1810 and 1815 and then posthumously in the Musée de L'Homme in Paris until 1985. Julia Pastrana (1834–1860), a Mexican woman displayed by her husband as an "ape-woman" in a touring "freak show" and discussed in *The Lancet,* provides further evidence of how the human-animal boundary was used to marginalize non-Caucasian women.[13]

Within the primate folklore tradition, the myth of consensual or nonconsensual sexual relationships between women (particularly African women) and primates was common. The sexual beast depicted was invariably a gorilla, which already by 1860 "embodied the essence of bestiality."[14] Male, hypersexual, and monstrous, the Victorian gorilla was both distant from humanity and fearfully close. Paul Belloni Du Chaillu played a significant role in creating this mythical beast. As the first Westerner to observe gorillas extensively in the wild, his dramatic traveler's account *Explorations and Adventures in Equatorial Africa* proved popular. In words and images, Du Chaillu painted a terrifying primate portrait. The following excerpt captures one of apparently several aggressive encounters: "Now he [the gorilla] was not twelve yards off. I could see plainly the ferocious face of the monstrous ape. It was working with rage; his huge teeth . . . [and] skin of the forehead . . . gave a truly devilish expression to the hideous face."[15] Later in the same chapter, he provided an image of a gorilla standing over an African hunter with his teeth bared and a bent gun in his hands (see figure 5). Not only a beast, here the gorilla was condemned as a killer.

Our Ape Other: Furious Beast or Civilized Companion?

Between 1860 and 1930, the primate stories written by or about explorers and hunter-collectors echoed in books, magazines, and films. Attracted by the growing financial reward of killing and stuffing primate specimens for museum displays or simply writing about primate encounters, these hunters and explorers were among the very few Westerners who actually saw apes in the wild, and the public accepted their accounts as an authoritative record of primates' appearance and behavior. Through popular periodicals, books, and early natural history films, intrepid adventurers became the authoritative source on information about how wild primates looked and behaved. Hunter-collectors like Ben Burbridge published adventure stories accompanied by images of gorillas that were far larger than life. Accompanying sketches showed hairy beasts dwarfing the human observer on the

Figure 5. "Death of My Hunter" opposite page 342 in Paul Du Chaillu, *Explorations and Adventures in Equatorial Africa* (New York: Harper and Brothers, Publishers, 1862).

magazine's front covers and inner pages. Similar amplification of apes' size took place on the big screen in films like Martin and Osa Johnson's 1932 *Congorilla: Adventures with the Big Apes and Little People of Africa*.[16]

At the heart of our enduring fascination with nonhuman primates was a belief that our closest animal kin could provide an authentic window into human nature and shed light on the drives associated with our "inner beast." The Depression-era blockbuster movie *King Kong* perhaps best embodied the mixed feelings awakened by gazing upon our fellow primates.[17] For members of the audience in 1933, Kong's ascent of the Empire State Building certainly evoked fear and compassion, but it also conjured questions that struck at the very heart of what it meant to be human: Can he feel love? Is he so different from us? What were we like in our natural form? How did civilization alter us? Does our core animal nature endure?

The commonality between humans and other primates that was depicted in popular publications and films also served to remind the public of apes' Otherness. Exaggeration of their size and overly dramatic depictions of primate behavior created a stereotyped fiction. Gorillas especially were shown in adventure magazines, best-selling books, and in films as several times bigger than their actual size and with a host of behaviors that stressed their lack of civilized virtue. Kong's character played on both sides of the issue. His depiction in some ways led the audience to a sympathetic empathy toward his emotional and moral capacities. At the same time, he was an enormous, brutish animal driven by his lust for a white woman. Thus, *King Kong* took the widely held cultural assumption that gorillas raped women, and it amplified the racial discourse by choosing a white woman, treated with tenderness, as the target of his desires.[18]

The trope of the aggressive ape was inevitably applied to the individual real gorillas who arrived in the United States in the 1920s and 1930s as pets, public spectacles, and scientific specimens. Buddy, for example, had spent time with missionaries in Cameroon and was later sold to a ship's captain. After traveling to the United States in 1932, Buddy sustained acid burns to his face as a result of an attack by a sailor while in port in Boston. This disfigurement only aided Buddy's final owners, the Ringling Bros and Barnum & Bailey Circus, who purchased him in 1937.[19] The circus' promotional campaign depicted Buddy, under his new name "Gargantua," as a giant and terrifying creature (see figure 6). "Gargantua the Great" was toured across the country and quickly became a household name.[20] His fame continued well after his death in 1949, with the authors of a 1977 account of Gargantua's life claiming

Figure 6. Strobridge Lithograph Co. "Ringling Bros. and Barnum & Bailey: Gargantua the Great," no date. (John and Mable Ringling Museum of Art Tibbals Collection)

that forty million people had seen him.[21] As recently as 2005, Gargantua was still attracting popular attention in the form of an exhibition of his skeleton and detailed life history at the Peabody Museum of Natural History at Yale University.

Other great apes attracted less attention than the gorilla, but a good many assumptions about the appearance, temperament, and intellect of primates accumulated nonetheless. As early as 1739, the chimpanzee was depicted as the most amiable of the primates, with a female chimpanzee being described as "having so many good qualities and perfections, that it is difficult to determine where to begin to relate."[22] These "little bushmen" were commonly likened to savages or children. In 1891, the British psychologist C. Lloyd Morgan described a primate, most likely a bonobo, that was held captive at a zoo in remarkably civil terms, highlighting her ability to sing and delicately sip beef tea (see figure 7). "Cousin Sarah," as he called her, was destined to become more savage with age, however. "The baby ape is much more like a human child than an old gorilla. . . . The development of

Figure 7. "Cousin Sarah" drinking tea from C. Lloyd Morgan, *Animal Sketches* (1891).

the savage brute-nature is accompanied by the development of a fierce and savage aspect." Despite the ever-increasing gap between the character of apes and humans, "the structure-chasm, in essential points, does not widen to anything like the same extent."[23] For Morgan, the chimpanzee and bonobo were physically very humanlike, but in terms of morality and intellect there was much that distinguished them from humans. Observations of a handful of apes in zoos formed the basis of this conclusion, which were printed in a children's book with the goal of encouraging youthful appreciation of the animal world and in turn faith in a divine Creator. Now known for developing Morgan's Canon, which declares the explanation for an animal's behavior that involves the simplest psychological abilities is the most scientific one, Morgan applied no such scientific rigor to his anecdotal descriptions of apes.[24]

For Americans in the early twentieth century who wished to see a chimpanzee for themselves, Richard Lynch Garner had a traveling show in which he appeared to have a conversation with a chimp named Susie. Garner was an amateur and scientific outsider; nevertheless, he published a book about primate communication in 1892 and proceeded to briefly rub shoulders with members of the American scientific community. Garner's *The Speech of Monkeys*, his tour with Susie, and his travels to Gabon where he used a phonograph to study primate communication attracted extensive press coverage at the turn of the twentieth century. Although popular with lay readers, members of America's scientific community were critical of Garner's work, with C. Lloyd Morgan describing *The Speech of Monkeys* as "anecdotal" and only suitable for "parlor tea-table" discussion.[25] It was not long until articles appeared disputing how much time Garner had actually spent in the Gabonese jungle, which left his scientific reputation permanently tarnished.[26]

Whereas Morgan and Garner wrote about primate behavior under the guise of scientific authority, others wrote fictional and playful tales that sprung from brief observations of captive primates. In *Joe the Chimpanzee and Other Stories* (1899), an unidentified female American author opened the text by claiming that she visited a chimp at the zoological gardens in Regents Park in London, England. The chimp was able to shake hands, understand questions, and light a fire "as cleverly as any Yankee boy [she] ever saw." When asked to read the *Times*, Joe "settled himself into the exact position of the comfortable English gentleman."[27] At the end of her visit, Joe apparently "took his hat, cane, and heavy wrap, and escorted us to the great door of the monkey house, shaking our hands as we bade him good-bye."[28] Although clearly a story that went beyond the realms of reality, *Joe the Chimpanzee* is similar to Morgan and Garner's accounts in centering the chimpanzee's apparent ability to socialize, communicate, and entertain.

Depictions of the chimpanzee as relatively civilized, friendly, and nonthreatening are representative of popular accounts of the ape during the late nineteenth and early twentieth centuries. Whereas the gorilla was cast as the cultural icon Kong, the chimpanzee inspired a softer literary representation in the form of the hit series of books starring Tarzan. Beginning in 1912 Edgar Rice Burroughs entertained the American public with tales of an ape-man in the jungle. Simultaneously a "fellow ape" and distinct for being "Man," Tarzan was at once a primitive killer of the jungle and a man able to attract the love of a white woman driven by her own "primeval nature."[29] On the page, a monkey occasionally accompanied Tarzan but beginning in the 1930s a chimpanzee called Cheeta emerged as his companion on the big screen, serving to remind the viewer of Tarzan's jungle origins.

Similar to chimpanzees, orangutans were seen as relatively benign creatures in scientific works. Commonly referred to as the "man of the woods," drawings of orangutans appeared in works like Edward Griffith's *The Animal Kingdom* (1827) and Edward Donovan's *The Naturalist Repository* (1824).[30] In the nineteenth century, Alfred Russel Wallace was among the most authoritative scientists who discussed orangutans, and he dedicated a chapter to the ape in his 1869 book *Malay Archipelago*. Wallace observed orangutans' locomotion and diet in the wild and shot and measured seventeen specimens, some of which were sent to museums in England. As in many hunting narratives of the period, Wallace described the kill in graphic detail. After Wallace had fired two shots, one orangutan male "lay groaning and panting, while we stood close round, expecting every breath to be his last. Suddenly, however, by a violent effort he raised himself up. . . . Another shot through the back caused him to fall down dead. A flattened bullet was found in his tongue, having entered the lower part of the abdomen and completely traversed the body."[31] With apparent ease, Wallace hunted these apes while keeping one infant, left orphaned after Wallace shot his mother, as a pet. He fed the primate rice water, bathed and brushed it, and briefly provided it with a substitute mother in the form of a buddle of buffalo skin.[32] Despite these efforts, the young orangutan died after a few months in Wallace's care.

The vast majority of the popular accounts of primates featured gorillas, chimpanzees, or orangutans. Other species, such as the gibbon or baboon, attracted far less attention. The gibbon's occasional bipedalism was highlighted in a few illustrations in the nineteenth century, but prior to the 1930s when the gibbon began to be identified as evolutionarily closer to humans than once thought, few things were written about it. The baboon also escaped the limelight. When it was described, it had none of the appeal or fascination of its great ape cousins. In 1940, William Atherton Dupuy, who wrote books on animals and insects and did some work for the U.S. Department of the Interior, characterized the baboon as "the lowest order of monkeys" and "a depraved and unlovely creature, full of cunning and malice, and given to blind rages of passion."[33]

Regardless of the species, observations of primates published in the popular press were frequently confined to captive specimens in zoos or circuses. When sightings were made in the wild, the observer was a hunter-adventurer or amateur, with the exception of Wallace, whose few pages of description of orangutans more closely resembled a hunting narrative than a scientific treatise. Their observations were at best anecdotal and at worst completely fabricated. Some authors like R. L. Garner became shrouded in controversy, while others, among them Edgar Rice Burroughs, built literary empires on the back of the public's primate fascination. Together, the slew of writings and images in books, magazines, and movies taught the American public about how primates supposedly looked and behaved.

Distinguishing Primate Science from Primate Folklore

When attempting to conduct the first scientific field studies of apes, early twentieth-century primate researchers had to navigate around the dominant cultural assumptions about primate biology and behavior that had long been engrained in the American consciousness. Imagine Harold Bingham's shock when in 1929, as a postdoc with American psychobiologist Robert Mearns Yerkes, he attempted to conduct the first scientific study of gorillas in the wild only to find them so shy that direct observation was impossible. The ferocious beast long seen on the pages of books was nowhere to be found in the jungles of what was then called the Belgian Congo. Instead, Bingham discovered an incredibly reticent animal, and he had to settle for examinations of gorilla scat and nests.

Although terrifying and dramatic images of gorillas had persisted for nearly a century, by the late 1940s scientists realized that gorillas did not actually resemble the caricature of aggression and sexual impulse that had dominated the public's perception of them since the 1860s. Studies of captive gorillas and Bingham's failed attempt to study them in the wild had led to the realization that gorillas are actually incredibly shy creatures, so shy in fact that researchers feared that observing them in the wild might be impossible. Although remnants of their aggressive reputation certainly remained, as reflected in lingering memories of Gargantua the Great, gorillas were slowly being reframed. E. G. Boulenger, curator of lower vertebrates in the gardens of the Zoological Society in New York, for example, included a chapter on apes and monkeys in his 1947 book *Wild Life the World Over*. He corrected the claims of "early travelers" who had described gorilla vocalizations as a "demoniacal cry," pointing out that "in reality its voice somewhat resembles the low bark of a dog" and that "those entitled to do so give the ape a good character, describing it as normally pacific, sometimes even friendly, as far as its relations with man are concerned."[34] Nevertheless, recasting the gorilla as an elusive and solitary figure was a slow process due to the lack of actual field observations of gorillas, with a complete reframing of the gorilla having to wait until George Schaller's successful observations of natural gorilla behavior in the 1950s. Schaller's research, in combination with Dian Fossey's later study of the same gorilla group, attracted sufficient popular attention to rehabilitate the public's image of gorillas. Despite these successes, the dramatic and oversized Kong was still resonant enough to create a blockbuster film once more in 2005.[35]

Other primates, like howler monkeys and gibbons, held their own surprises for researchers. Clarence Ray Carpenter, another of Yerkes's postdoctoral fellows, discovered that both howlers and gibbons were capable of far more complex communication than previously thought. Their expressions proved to be much more than merely uncontrolled, emotional, vocal outbursts. Individual calls coordinated social behaviors, searched for lost young, and warned of approaching predators. Carpenter's early studies also generated a host of basic biological information about primates' size, menstruation patterns, social structures, and sexual behaviors, much of which contradicted the popular assumptions about primate biology that had long been propagated by the fictitious accounts that had dominated primate folklore.

Framing new descriptions of primates and their behaviors as an effort to separate fact from fiction was nothing new. Tyson's 1699 work on the orangutan, for example, concluded with "A Philological Essay Concerning the Pygmies of the Ancients" in which he quite specifically sought to replace the musings of a poet with the facts of an anatomist:

> Having had the opportunity of dissecting this remarkable creature, which not only in the outward shape of the body, but likewise in the structure of many of the inward parts, so nearly resembles a Man . . . as it suggested the thought to me, whether this sort of animal, might not give the foundation to the stories of the Pygmies? and afford an occasion not only to the poets, but historians too, of inventing the many fables . . . that are transmitted down to us concerning them? I must confess, I could never before entertain any other opinion about them, but that the whole was a fiction.[36]

A similar call for replacing fables with facts appeared in 1861 when the periodical *Temple Bar* published, "With Mr. Gorilla's Compliments." The author was anonymous, but the article clearly reinforced Sir John Edward Gray's public critique of Du Chaillu. Gray, head of the British Museum's zoological department, accused Du Chaillu of misidentifying primates and drawing gorillas from previously published images rather than from direct observations in the wild, as he claimed.[37] The author presented the perspective of a gorilla, claiming that "my family [has] been grossly scandalized for many years by persons none of whom can boast the honour of even a personal acquaintance with us, and by far the majority of those who talk so very loudly about our affairs . . . [have never] set eyes on us."[38] Despite Gray's efforts, and those of the anonymous author who hoped that the public would "prefer truth to miracles," the public's infatuation with primate fictions was far from over.[39]

More than three centuries of primate folklore and the public's long-standing fascination with primates was both a blessing and a curse for early primate researchers. The public was clearly fascinated by what other primates looked like, how they behaved, and how they were similar and different to humans. However, the public also had inaccurate notions about primates, especially the great apes that proved difficult to overcome. The image of the gorilla as an enormous physical and sexual aggressor reached its height in the 1930s but continues to endure today. Similarly, connections between chimpanzees and tea drinking motivated zoos to hold chimp tea parties well into the twentieth century.[40]

Perhaps even more important to the early scientists who sought to study wild primate behavior, the public and their fellow scientists had preconceived conceptions of the type of person who studied primates and the methods they used to gain their information. Amateurs and hunters who sought the limelight were the ones who wrote about primates. In almost all cases they were men. Those few who saw, or claimed to see, primates in the wild were adventurers, enduring difficult environments and encountering apparently dangerous and primitive natives in pursuit of a glimpse of an ape and hopefully a clean shot. All the accounts failed to demonstrate—or even strive for—any rigorous scientific methodology, which resulted in exaggerated, anecdotal observations and in some cases outright lies. Some aspects of this reputation may have been alluring to early primate researchers seeking adventures in far-off lands, but no aspiring scientist wanted to be identified with sloppy observation.

The first generation of researchers who sought to conduct scientific studies of primates in the wild had reasonable hopes. They expected to be able to see the primates and record their locomotion, diet, and sexual and social behaviors. In short, they expected to do a much better job than the amateurs that had come before them. As the next chapter will demonstrate, obstacles and disappointment characterized the first stage in the quest to see primates scientifically. In the 1920s, a select group of American researchers set off to gain knowledge of wild primate behavior with the goal of replacing traveler's tales with systematic field science. At first, they believed seeing primates in the wild would be as easy as finding "rabbit tracks in fresh snow." Unfortunately, the first attempts to scientifically study primates in the wild bore more resemblance to searching for a needle in a haystack. These difficulties forced scientists to collaborate with individuals with experience seeing and keeping primates—the very amateurs from whom they sought to distance themselves—while simultaneously distinguishing their science from the stories of their predecessors.

2

Venturing out of the Lab and into the Wild

Early Primate Field Studies

By the 1920s over two centuries of mythmaking about nonhuman primates had created a folklore consisting of exaggerated stories of primate behavior and inaccurate images of their physical form. Literary figures and hunter-collectors created the stories that formed the heart of this folklore. At various points in this history, naturalists like Edward Tyson and John Edward Grey attempted to replace these primate stories with scientific observations of primates in the wild. These attempts were thwarted because the individuals who were actually out in the wilderness seeing, or at least claiming to see, primates in the early twentieth century were adventurers and amateurs, not scientists.

In contrast to the adventurers who went out into the wild, members of the scientific community examined primate skins and skeletons from the safety and comfort of their university offices and laboratories. Their work yielded important evidence for discussions of classification, but it was of little help when trying to glean how the animals behaved in their own habitat. Venturing out of these academic settings to observe apes and monkeys in zoological parks partially filled this void, but it was not enough. Seeing and understanding how primates functioned in social groups and in interaction with their environments required actually seeing them in the wild. Performing this fieldwork would necessitate a new generation of scientists prepared to act more like adventurers, taking on the physical and methodological challenges of making prolonged observations of primates in their native habitats. Preparing scientists for fieldwork and generating basic knowledge of primate biology, however, demanded preserving, rather than completely severing, connections between amateurs and their primate subjects. In short, early primate researchers needed the expertise of amateurs and thus generated a range of collaborations with individuals outside the formal scientific community.

One of the most influential individuals in the history of primate studies was American psychobiologist Robert Mearns Yerkes, who sought to bring primate folklore under the control of science by making scientists, not adventurers, the authorities on primate behavior. The journey away from primate folklore and toward primate science required building a body of knowledge about basic primate biology, reproduction, cognition, and social behavior from the ground up. Achieving this goal involved both generating the funding necessary to get scientists into the field and developing the methodologies needed to enable them to make prolonged observations of natural primate behavior once they got there.

Yerkes secured funding for the first scientific field studies of primate behavior and created the institutional framework to support the series of postdoctoral fellows employed to carry out these early studies. But behind any founder of a new discipline are a range of collaborators and assistants, who in various ways assist in the development of a discipline or subdiscipline. This was particularly true in the case of primatology. Specimens needed to be accessed, and methods and sites needed to be created. For assistance in these matters, Yerkes turned to his wife, Ada Yerkes, and two scientific outsiders—Rosalia Abreu, a wealthy Cuban woman who owned captive chimpanzees, and Ben Burbridge, a hunter-collector with access to a mountain gorilla.

With the aid of Ada, Abreu, and Burbridge, Yerkes's primate research initially focused on laboratory environments and outdoor spaces ranging from private estates to backyards and circus quarters. Here Yerkes began to accumulate basic information about primate biology, reproduction, and cognition that was still unknown in the 1920s: What did chimpanzees eat? How could chimpanzees be kept alive and stimulated to reproduce in captivity? What was the intellectual capacity of chimpanzees, gorillas, or orangutans?

Although fruitful in respect to studying primate biology and intellect, and in creating an institutional framework for scientifically studying primates, these early captive studies were incapable of revealing natural primate behavior. Only in a primates' native habitat could researchers see authentic social behaviors between members of a primate group and interactions with other species. To access this knowledge, Yerkes hired three postdoctoral fellows to conduct the first scientific field studies of primate behavior. He had high hopes for these expeditions but it quickly became apparent that making systematic observations of primates would not be an easy task, especially because any claims to scientific authority would need to be strong enough to overcome the well-established cultural associations between adventure, amateurs, and apes.

Robert Mearns Yerkes's Early Laboratory Studies of Primates

Trained as a psychologist, Yerkes received his PhD from Harvard in 1902 for his study of the sensory reactions and nervous physiology of jellyfish and went on to work at Harvard as an instructor in comparative psychology until 1917.[1] During his early professional career, his research included the sensory abilities of frogs, inheritance of dancing behavior in mice and "savageness" in rats and the motor functions of turtles.[2] Nine years after receiving his PhD, Yerkes wrote *An Introduction to Psychology*, a textbook for psychology courses.[3] It marked the beginning of a career characterized by prolific publication, with many of Yerkes's texts making significant contributions to psychology. However, his research was not limited to any one discipline. Rather, he crossed disciplinary boundaries, migrating into psychology, biology, anthropology, psychiatry and what would eventually become primatology. Yerkes used the term *psychobiology* to capture the transdisciplinary nature of his research.[4]

Yerkes understood his transition from studying mice to primates as an extension of American behaviorism. Usually associated with the study of rats or mice, Yerkes conceptualized his primate studies as taking fundamental tenets of behaviorism and applying them to primates. Thus, in 1916, at the dawn of his career-long preoccupation with primate behavior, he envisioned a laboratory for primate studies that would explore "the behavioristic form of the perennial questions: Do these animals think, do they reason, and if so, what is the nature of these processes as indicated by the characteristics of their adaptive behavior?"[5] While pursuing these questions, Yerkes developed a multiple-choice apparatus, box stacking problems, and a host of other food reward techniques for the study of primate intelligence.

Yerkes led the institutional, intellectual, and social development of the emerging discipline of primatology. He established the Yale Laboratories of Primate Biology in New Haven in 1925, followed five years later by the Anthropoid Experiment Station in Orange Park, Florida. Yerkes also attempted to incorporate the study of primates into animal behavior studies more widely through extensive contact with field and lab-based biologists.[6] His papers reveal a high level of professional etiquette, replying promptly to inquiries, providing advice to many colleagues, and being an active member of a diverse range of committees and conferences. As a result, Yerkes was well regarded both in biology and psychology. These connections, and others within animal behavior studies, ornithology, and zoological parks, aided him in expanding both the intellectual and social base of this emerging discipline. For example, his participation in the National Research Council from 1919–24, specifically his work with the Committee for Research of Problems of Sex (CRPS), helped secure a reliable source of funding for primate studies from the late 1920s through the 1930s, including a $23,000 CRPS grant to support his own laboratory at Yale from 1925–28.[7]

Yerkes's involvement in the CRPS, a committee that sought to bring understandings of sexuality under the control of science, dovetailed nicely with his vision for a primate laboratory. This intersection ensured early primate studies would be characterized by a preoccupation with sexual behaviors, including masturbation, copulation, and sex drive. Ultimately, through the sites he developed for working with primates, his training of postdoctoral fellows, and his creation of funding opportunities for primate studies, Yerkes helped build the foundation upon which the discipline now known as primatology was built.

Like many researchers in the early twentieth century, Yerkes was one half of a "creative couple."[8] In 1905, he married Ada Watterson, with whom he coauthored the 1929 book *The Great Apes* (see figure 8), the publication for which Ada is best known.[9] Ada was educated at Smith College and received her PhD from Barnard, where she later taught botany. She established plants and planned the landscaping at both the New Haven and Orange Park laboratories.[10] Ada continued to make significant contributions to her husband's work, receiving regular acknowledgments from Yerkes for her help with various books and coauthoring a number of articles. Specifically, she coauthored papers about the relationship between psychobiology and psychiatry, between studies of primates and "the conscious control of human life," and chimpanzees' fear responses to foreign objects. Furthermore, she

Figure 8. Robert and Ada Yerkes. (From Robert Mearns Yerkes Papers [MS 569], Manuscripts and Archives, Yale University Library, box 137, folder 2291)

coauthored two works serving primarily as reviews of existing literature: *The Great Apes* mentioned above and a lengthy article about primate social behavior.[11] Yerkes's respect for and collaborations with his wife undoubtedly played a role in his professional relationships with other women by increasing his awareness of the contributions women could make to primate research. Ada was also socially active in animal behavior studies as she and her husband hosted a number of investigators working with primates and other organisms at their summer home, Franklin Farm, New Hampshire, which served as a field station. The Yerkes had two children, Roberta and David. At times readers catch a glimpse of this family life in the midst of Yerkes's scientific publications. For example, some of Yerkes's work includes sketches of primates made by his children. David Yerkes also worked briefly with primates, publishing a paper with his father about a study of memory conducted at their summer home.[12]

Much of Yerkes's work with primates took place first at the Harvard Psychology Laboratory and later at the Yale Laboratories of Primate Biology, where he examined fundamental characteristics of primates such as ways to maintain their health and reproduction and their forms of social organization. He also chronicled their drug addiction as a result of experiments.[13] Despite being situated in a laboratory, his research was rarely, if ever, conducted in the sterile, isolated place that word typically brings to mind. Although aspects of the sites used by Yerkes were reminiscent of traditional laboratory settings, the use of outdoor enclosures and alternative spaces, such as a barn at Harvard, ensured that much of Yerkes's work involved what he referred to as "messy maintenance."

Yerkes's Harvard colleagues criticized the extensive housekeeping involved in his research.[14] Nevertheless, a significant amount of the work conducted at Harvard, Yale, and later at the Anthropoid Experiment Station in Florida involved such basic practical concerns as how to feed and house the primates. Such concerns were paramount, for without a healthy, self-reproducing colony any research projects would be short lived. Indeed, Yerkes identified "promotion, housekeeping and research" as his three main activities, adding that he had: "placed them in order of priority and logical dependence. Without the promotion of my idea and its embodiment in steel, concrete, flesh and blood; without financial support and the energies of investigators as means, progress would have been impossible."[15] In fact, it was the relationships Yerkes formed while rearing primates, rather than his research accomplishments, that he found most rewarding: "The task may become irksome and discouraging if one realizes that little credit is likely to come from the solution of housekeeping problems. . . . But to one who enjoys and has a flair for keeping animals contented and in good condition . . . the 'housekeeping' of an anthropoid breeding colony may bring deeply satisfying rewards. Such an experience has been mine as dividend over and above the rewards of discovery through observation and problem solution."[16] This quote echoes the rhetoric surrounding domestic housekeeping and home economics of the period. Unlike the average housewife, however, Yerkes's work revolved around caring for a captive primate family.[17]

Like many women during the early twentieth century, Yerkes found "deeply satisfying rewards" in domestic tasks. Those around him also conceptualized his work in terms of a family unit. Richard Sparks, a correspondent of Yerkes and an advocate for gorilla welfare in the 1930s, responded to Yerkes's announcement that his primate colony had grown to some twenty-eight chimpanzees by 1931 with this: "that is certainly some family to care for, but I know you have pleasure in every day of your work with them."[18]

In addition to cleaning and care, the establishment and maintenance of primate colonies required money. In search of funding, Yerkes embarked on extensive promotion of the concept of primate laboratories at New Haven and Orange Park to philanthropic associations, eventually gaining a four-year $500,000 grant from the Rockefeller Foundation in 1929. As chronicled in his autobiography, gaining such financial support took a great deal of social energy: "For at least ten years before the financial problem was solved in 1929, I had been using every opportunity to acquaint the officers of foundations, and especially those of the Carnegie and Rockefeller organizations, with the research values of primates.... This spadework brought me into pleasant contact with several administratively wise and disinterested men. It had the advantage of making me known to them personally and of awakening their interest in a novel proposal for biological research."[19] The friendly relationship that ultimately developed between Yerkes and the Rockefeller Foundation is demonstrated by an amusing exchange of letters written from the perspective of the primates at New Haven—William, Dwina, Pan, and Wendy—to the "Rockefeller Philanthropoids" when the breeding station at Orange Park was established. The Yale anthropoids received a reply from the institution expressing that it had been hoped that, rather than being transferred to the station at Orange Park, the chimps would come and "assume the administration of the Foundation."[20] Clearly, Yerkes had accumulated a significant amount of social and financial capital by 1929. He had also successfully added primates to the list of animals seen as valuable subjects for science. This financial and intellectual support laid the foundation for early laboratory-based studies of primate biology and intelligence.

Yerkes Steps outside the Laboratory

Despite his many contributions to laboratory studies of primates, Yerkes was a keen advocate of studying animals outside of the lab from early in his career. For example, as early as 1912, prior to the construction of the Yale Primate Laboratory and while Yerkes was still at Harvard, he established a non-laboratory site for the study of primates in his private summer home. Over the course of two years, Robert and Ada Yerkes purchased two farm buildings on a one hundred and fifty-acre area of semi-wooded land in Franklin, New Hampshire. Referred to as "Franklin Farm" in Yerkes's casual communications and guest book and identified in his published works as the "Franklin Field Station," this site was initially used by scientists associated with Harvard, and later with Yale, for the study of animals ranging from earthworms and crows to primates.

Yerkes's commitment to the importance of ensuring that "every student" be "familiar with the objects of his interest in nature as well as in the laboratory" motivated his purchase of the property.[21] He intended such exposure to ensure that the "tendency toward the acquisition of a narrow minded attitude" commonly associated with the laboratory was "counteracted" by training students "to become enthusiastic and reliable naturalists as well as skilled experimentalists."[22] Furthermore, Yerkes argued that the Franklin Field Station would provide an important supplemental site to Harvard Psychology Laboratories. Specifically, he argued that the farm would increase efficiency by enabling the continuation of research, and thus the ongoing development of skills, during the summer months. This increase in efficiency was closely connected with the opportunity to enjoy a break from school and city life by retreating to the countryside. He wrote: "The Franklin Field Station, it is hoped, will result in the saving of considerable time to certain investigators, since there it should be possible to continue work uninterruptedly throughout the summer, while at the same time the investigator may profit by some change from city to country and the chance to combine experimental and naturalistic studies in animal behavior with the recreations of a mountainous country."[23] Such combination of work and play was similar to marine biological laboratories like those at Friday Harbor, Washington, and Woods Hole, Massachusetts.[24] Like these sites, the farm welcomed the families of scientists. In this respect, Franklin Farm is also reminiscent of British country houses and lodges of the Victorian era where domestic and laboratory spaces overlapped, and in turn wives and children were interwoven and active participants in the science conducted there.[25]

Yerkes studied Chim and Panzee, who he believed to be two chimpanzees although Chim is now known to have been a bonobo, at Franklin Farm for eight weeks during the late summer and early fall of 1923, observing their adaptability, affective expression, and evidences of insight. The study was then completed in the Yerkes's winter residence in Washington, D.C., where Chim and Panzee lived in a cage with access to an open-air porch while Yerkes worked for the National Research Council (NRC). Yerkes's observations focused on physical and mental traits of the two chimps and contrasted the animals' temperaments and reactions to the urban setting of D.C. with the rural surroundings at Franklin Farm. The primates' fear of cows, their nest building in the surrounding birch trees, and eating habits at the dinner table were all described, although the level of detail pales in comparison to the depth seen in some of Yerkes's later work in similar settings. Yerkes believed that such preliminary studies remained valuable due to the fact that little was known about primates' life histories or behaviors and because he remained committed to the potential of primates to reveal knowledge about the development of human societies.[26] Yerkes also took some pains to delineate his observations from those of travelers and other amateurs who lacked scientific training in the study of animal behavior with his use of practices such as box stacking experiments to study adaptability and insight serving to highlight the scientific nature of his studies despite their occasionally being located in his living room.

Yerkes's attempts to study primates outside of the lab extended beyond the domestic setting of Franklin Farm to Cuba and the home of Rosalia Abreu (1862–1930). Hailed as "a Cuban patriot" who "devoted much of her great wealth to [Cuba's] fight for freedom," Abreu was famous throughout Cuba and beyond for her large chimpanzee colony (see figure 9). Like many who devoted their lives to animal study, Abreu was infatuated with unusual pets from childhood, a passion encouraged by her father. By the turn of the century Abreu had apparently amassed the world's most extensive and healthy captive primate colony, which she housed on her estate, Quinta Palatino, in Havana. Newspapers highlighted Abreu's extreme wealth and her dislike of people, with one headline in 1930 identifying her as "one of the strangest women of our time."[27] The reporter's interest in Abreu's personality was shared by many, including Yerkes, who dedicated many pages of his 1925 book *Almost Human* to a discussion of Abreu's personality traits and intense affection for her subjects.[28] Abreu's mystique also spread throughout Havana, where her estate was referred to as "la finca de los monos," or "the monkey farm."[29] After her death, an article by the *New York Times* highlighted Abreu's rise to "scientific prominence" as a result of her work with Yerkes and Yale University, a relationship that endured after her death with the donation of her primates to Yerkes's laboratories.[30]

Abreu's estate consisted of a large area of well-kept, beautiful gardens within which were a number of small cages. Although the grounds contained a variety of large mammals, including an elephant and several bears, primates occupied the majority of space in both Abreu's estate and in her heart. Anumá, the first chimpanzee to be born in captivity, was born during the summer of 1915 at Abreu's estate. Yerkes was aware of Anumá, and the achievement his birth represented, and it was this news, combined with the encouragement of William Morton Wheeler of Harvard University, that initially pushed him to contact Abreu.[31] What followed was a number of letters based on a shared curiosity concerning primates and their behaviors and Yerkes's visit to the colony in the summer of 1924, a trip funded by the Carnegie Institution. Ultimately, Abreu's observations of the behavior of her pets formed a substantial part of Yerkes's book *Almost Human*.

Almost Human drew upon the experiences and observations of Yerkes's own animals in New England, as well as those of several private individuals, including Alyse Cunningham, who kept gorillas as pets, and Nadia Kohts,

Figure 9. Rosalia Abreu is the woman on the far right. Her son, Pierre, is in the middle. (Photo kindly provided by Rosalia Abreu's great-grandson, Eric del Junco)

who was involved in primate studies in Russia.[32] The vast majority of the volume, however, was based on the experiences Abreu had gained from keeping primates for nearly twenty years and on Yerkes's own observations while visiting Quinta Palatino.[33] Yerkes described the physical characteristics of the colony, the cages, surrounding trees, and species in some detail. More significant, Yerkes directly quoted Abreu's opinions concerning primate behavior, only rarely pointing out where they conflicted with his own views. He also stressed Abreu's abilities to handle the primates and maintain both their health and their ability to reproduce. Two chapters were dedicated to what could be learned from the cages and conditions of Abreu's colony in terms of maintaining healthy primates in captivity, including information about the flooring used, surrounding landscape, human and intraspecies interactions, temperature conditions, and diet.[34]

Beyond basic primate husbandry, Yerkes utilized private estates and collaborations with scientific outsiders to gain knowledge of primate's intellectual capabilities, specifically those of the mountain gorilla Congo. Named after the country where she was captured, Congo was brought to the United States in 1926 by hunter-collector Ben Burbridge (see figure 10). As the first female of her kind to come to the United States, Congo was greeted by publicity. Rather than stories of aggression and sexuality, which were common in depictions of gorillas, it was Congo's womanhood and human-like nature that dominated these popular accounts: "The first thing one is apt to notice about Congo is her eyes. . . . The face is ape, but the eyes are indubitably human. . . . The hands and feet . . . are so nearly human that they take one's breath."[35]

Burbridge enjoyed hunting with "gun and camera" and published a series of articles about his collecting and filming of live gorillas in what was then the Belgium Congo in the natural history periodical *Forest and Stream*, one of many such periodicals established in the late nineteenth century appealing to the hunting community.[36] Burbridge argued that, instead of being stereotyped as a particularly aggressive primate, the gorilla should be seen as "remarkably individualized."[37] He emphasized their individuality by providing descriptions of his encounters with males, females, and young primates, some of whom exhibited aggressive behaviors while others were intensely curious about both Burbridge and his motion picture camera.[38] Nevertheless, it was the encounters with gorillas characterized by danger and aggression that he depicted most vividly in his articles and in the front cover illustrations that adorned the 1927 and 1928 issues of *Forest and Stream*.

Figure 10. Congo two weeks after capture with unidentified man. (Richard D. Sparks Papers, box 1, folder 13; courtesy of University of Arizona Libraries, Special Collections)

These observations and the gorillas he captured were in part intended as contributions to science, although Burbridge realized that "extensive study and experiment" would be necessary to understand the range of gorilla vocalizations and mental capabilities.[39] Burbridge went on to publish a book, *Gorilla: Tracking and Capturing the Ape-Man of Africa*, based on his experiences while trapping and capturing gorillas in the wild. Although the subject of its title, gorillas were detailed in only three chapters of the book, much of which duplicated the text used for his series in *Forest and Stream*, with one additional chapter dedicated to a discussion of "Miss Congo." The latter included an account of Yerkes's observations of Congo from Burbridge's perspective. Burbridge was obviously proud of his gorilla's capabilities and the role she had served for science.[40]

Yerkes's work with Burbridge represents one example of the relationships that frequently formed within animal behavior studies between hunter-collectors and members of the scientific community. Hunter-collectors commonly assisted scientists while in the field in finding primates, collecting samples, and obtaining photographs. For example, Burbridge supplied a gorilla for the Zoological Garden in Antwerp. In Yerkes's case, the interaction occurred on American soil after Burbridge had returned to the states with Congo. He agreed to let Yerkes study the gorilla, arranged for this research to take place on his brother's estate, where Congo was living at the time, and shared with Yerkes his own observations while trapping gorillas in the wild.[41]

Yerkes also gained information concerning methods of observation from Burbridge. Initially, Burbridge spent a few frustrating months attempting to observe the gorillas by imitating their calls and beating his chest. Eventually, Burbridge used the gorillas' curiosity and "poor mathematics" to successfully observe and film their behaviors. By this he meant that Burbridge and the Congolese who accompanied him would attempt to ensure that the gorillas saw them hide in the forest. The majority of the group would then leave the hideaway and walk away, "tricking" the gorillas into thinking that the whole group had left the vicinity when in fact Burbridge with his gun and camera and one or two Congolese would stay behind. They would then wait for the gorillas to enter the clearing, apparently often enticed by their curiosity about the motion picture camera.[42] At one point, the interest of a young gorilla in Burbridge's camera was likened to "a small boy peering over the fence at a ball game."[43] Burbridge's *Forest and Stream* articles described these techniques and thus Yerkes would have benefited from these experiences, either directly from Burbridge or from his publications. Yerkes certainly recognized Burbridge's contributions to the study of gorilla behavior and intelligence, describing Burbridge's knowledge of gorillas as incredibly useful and writing that "it is doubtful whether anyone knows the mountain gorilla more intimately or, for the purpose of hunter and photographer, more usefully than does Burbridge."[44] Yerkes's respect for the hardships Burbridge had endured in Africa certainly played a role in boosting his opinion of this hunter-collector. However, it is significant that Yerkes pointed out Burbridge's value specifically as a hunter and photographer but not for his collection of scientific knowledge.

Yerkes worked at "Shady Nook," the estate of Ben Burbridge's brother, James Burbridge, for six weeks during the winter of 1926 and then again in 1927 for eight weeks.[45] This research focused on Congo's mental abilities, studying adaptation as a sign of mental ability and involving many experimental techniques with which to measure Congo's intelligence. The practices used included box problems, such as testing to see if the primate could stack boxes to reach a food reward, multiple-choice experiment, location of buried food, and recognition of reflection in a mirror. When recording his experiments, Yerkes noted every detail, including the time of day when experiments started, the temperature, the size of the enclosure or the kind of outdoor space being utilized, and Congo's response to the experiment every time it was performed. Repetition was a central characteristic of Yerkes's study. For example, when examining Congo's abilities to recall information through delayed response experiments, Yerkes conducted thirty-nine trials. Shorthand notes were made during the experiments in the morning, transformed into longhand notes in the afternoon, and later supplemented with still and motion pictures. The final result was descriptions such as the following: "With Congo waiting at the post and everything arranged for observation, I let her see me place in the black box a quart cup of milk. As soon as the box was covered I started my stop watch and promptly took my place beside the release mechanism, where I observed and recorded her behavior during the five minute delay. . . . At the proper moment . . . I loosened the rope and, at once appreciating the measure of freedom thus given, she proceeded directly and rapidly to the black box and eagerly drank the milk."[46] Yerkes's thick description and use of repeated experiments highlight his combination of the kind of detailed observation commonly associated with natural history with scientific values commonly associated with the laboratory. Yerkes's belabored descriptions of his methods, characteristic of all his scientific publications, reflect his preoccupation with conducting a particular sort of science.

Yerkes's text and photographs give glimpses of Congo's experiences during the intelligence studies. As an experimental subject, Congo was frequently chained to the ground or controlled with a rope and pulley system to restrict her mobility. She was also punished at times, for example when guessing incorrectly using the multiple-choice apparatus, in the form of electric shocks.[47] Congo may also have been subjected to physical abuse earlier in her life while owned by Ben Burbridge, who was known to "break" gorillas with "his fist and whip." Congo's "gentle" nature spared her these forms of "extreme punishment" but it seems likely she was a victim of some degree of physical violence.[48]

For studies of primate intelligence, Congo was "potentially a scientific prize," providing Yerkes with the opportunity to study the mind of a gorilla, specifically the mental development of a gorilla over time, an opportunity for which he had waited all of his life.[49] Yerkes used this powerful statement to open his first of a trilogy of articles on "The Mind of a Gorilla" published in the scientific journal, *Genetic Psychology Monographs*. He pointed out that his study of Congo was "unique because it records the first mental measurements of a gorilla" and thus was a "pioneer contribution" to science.[50] This enthusiasm for the study of gorillas, especially the mountain gorilla about which little was known, was not peculiar to Yerkes.[51] When a group of researchers, including the American explorer and sculptor Carl Akeley, met with Yerkes to discuss the possibility of studying Congo each "made no attempt to conceal" their "eagerness for intimate acquaintance with the gorilla, one or many."[52]

Intimacy was a theme in most of Yerkes's studies of primates, with the first few pages of many of his publications focusing on the relationship between the primate subject and his owner and/or Yerkes himself. In the first "Mind of a Gorilla" article, Yerkes described Congo as a "gorilla child" cared for by both Mr. and Mrs. James Burbridge (see figure 11). Through detailed and emotive descriptions, Yerkes communicated the connection he felt with the subject of his study:

> When Mrs. Burbridge entered the cage to play with Congo I accompanied her, and she took the gorilla-child on her lap, stroking her heavy coat and talking to her the while gently and assuringly, I stood at one side waiting for Congo to make the first advance. Presently she reached out her left hand to me and when I grasped it she drew me to her. Then followed close visual and olfactory scrutiny of my hands, face, and hair. Satisfied, apparently, that I was worthy of her friendly confidence she put an arm about my neck. From that moment Congo and I were trusting friends.[53]

After these first few pages, Yerkes quickly reverts to a scientific style of writing, describing in detail all aspects of the gorilla's body, cage, and the various experiments conducted. However, he restated the importance of an intimate relationship between subject and experimenter in a later article, specifically citing the need for the primate's cooperation so that the experimenter may work with such a strong animal in safety.[54]

The safety of the animal was also a concern. Rather than risk the dangers posed by transporting Congo during the winter months from her home in Florida to Yerkes's laboratory at New Haven, Yerkes traveled to Jacksonville to conduct his study of Congo's mental abilities. Aware of the distinction between this outdoor environment and a traditional laboratory, Yerkes opted to embrace the nature of his surroundings rather than seek to construct some kind of laboratory for his experiments. He wrote: "My initial decision was to adapt myself and my research interests as far as practicable to the conditions at Shady Nook instead of attempting to convert it into a scientific establishment. . . . I therefore accepted out-of-doors at Shady Nook as my laboratory."[55] Clearly, Yerkes was aware that valuable information could be acquired outside a traditional laboratory setting. Reflecting on his work with Congo he concluded that he was "reasonably certain that had I worked more conventionally and in accord with laboratory practice I should have acquired less information and insight."[56] Yerkes made no attempt in his publications to conceal the out-of-doors nature of the study, with frequent references to the Burbridges and their home.

Despite the accomplishments of Yerkes's work with animals in domestic estates, his study of Congo revealed the limitations of semi-captive sites for studies of primate social behaviors. Restricted to the interactions between Congo and two pet dogs, with her human caregivers, and with a mirror, Yerkes made relatively superficial, albeit interesting, observations of Congo's social behavior. In 1926, Congo had played with the Burbridges' dogs, Betty the Airedale and Bobby the Bulldog, as companions. This changed when Yerkes returned in 1927. Betty was still at the estate, Bobby the Bulldog was no longer there but there was now a male mongrel. Congo was a year older and now engaging in "sex play" with Betty and the unnamed mongrel, which took the form of Congo "stepping astride" the mongrel male and "assum[ing] a male copulatory position and execut[ing] appropriate movements." This was followed by Congo lying down and bringing the dog onto her belly "in what had every appearance of sex embrace."[57]

Congo's use of dogs as "sex objects" was not dissimilar to observations of

Figure 11. Robert Mearns Yerkes with Congo. (From Robert Mearns Yerkes Papers [MS 569], Manuscripts and Archives, Yale University Library, box 131, folder 2238)

other captive animals while in puberty. The endurance of Congo's response to the social companion reflected back to her by a mirror, however, was quite unusual. Other monkeys and apes would initially search for the "original of the image" and then accept defeat. In contrast, "on the last day of observation ... [Congo] was as keenly interested in the companionship of the mirror as

Figure 12. Congo hugging mirror, 1927. (Clark University Monographs)

originally."[58] With a possessive attraction, Congo held onto her glass friend (see figure 12). Take for example this account of Yerkes's seventh repetition of the mirror experiment: "She mouthed and kissed the image, felt for it on the glass and behind the frame, turned the frame about and searched behind it, tried to carry it with her into the nest-room, and finally was unwilling to let me take it from the cage with me."[59]

For Yerkes, Congo's loneliness provided one possible explanation for her enduring confusion over her self-reflection. The other option was a "slowness of intelligent adaptation to a novel type of situation." Indeed, overall, Yerkes found Congo wanting of the kind of intelligence he had identified in chimpanzees. According to Yerkes, this relative mental deficiency was the probable reason for the gorilla being "outstripped . . . in the race for anthropoid supremacy."[60] In a harsh critique of Congo's mind, Yerkes wrote: "She is too much aloof from her environment, too little adventurous, or, in the scientific sense, inquiring, to readily and quickly discover solutions to novel problems and adapt herself to extraordinary environmental demands. . . . One can understand why the gorilla should be a disappearing race, and perhaps also why so little relatively is known about its mental traits, and so little sympathy exists between man and gorilla."[61] Such an excerpt echoes the sins of femininity: Congo is too aloof, too timid, and possesses too little intelligence.

The location of Yerkes's research shifted again in January and February of 1928, with further observations of Congo's mental development taking place in a circus environment. The circus may at first glimpse appear an "unscientific" site for research; however, the study continued to be supported by a combination of funds and equipment from the Laura Spelman Rockefeller Memorial, the Rockefeller Foundation, and the CRPS-NRC, and it also continued to involve the use of experimentalist practices.[62] By this time, a change in location was necessary because Congo had grown in size and aggression and thus was no longer owned by Ben Burbridge or housed at his brother's estate. Instead Congo was under the care of John Ringling and kept at the winter quarters of the Ringling and Barnum and Bailey Circus in Sarasota, Florida.[63] Yerkes expressed his gratitude for the assistance provided by Ringling and other members of the circus staff, especially in terms of learning how to handle and care for gorillas, in the opening page of the third and final part of his series of articles titled "The Mind of a Gorilla."[64] He was disappointed by the circus quarters, however, as a site for the continuation of his research.

Ideally, Yerkes's study of Congo's mental development would have benefited from continuity in terms of the site of research. Thus, the transfer of ownership of Congo from Ben Burbridge to John Ringling was disappointing. Fortunately, she adapted well to her new home on Ringling's estate, Ca'd'Zan, where she benefited from a "tropical appearance" and could see a fellow captive primate, a chimpanzee also kept on the grounds. Yerkes, however, elected not to study Congo in her new home and instead decided to conduct his research in the circus quarters. One of the reasons for Yerkes's decision was his belief that the circus location would provide "exceptional facilities ... for the construction and adaptation of apparatus and the control of experimental conditions."[65]

Despite some conveniences, such as equipment for experiment construction, the circus posed many challenges to Yerkes. The many distractions, including a high level of noise, drew Congo's attention away from the artificial problems laid before her: "Environmental conditions were novel, variable, and highly distracting, for elephants, camels, monkeys, horses, and men were in sight and most of them were moving about and creating more or less disturbance. Within a hundred yards a sawmill was in operation. Small wonder then if Congo was distracted, disturbed, and at times frightened."[66] At the same time, the circus offered opportunities to expand Yerkes's inquiry into the gorilla mind. Unlike Shady Nook, this new location enabled Congo's reaction to a wide range of animals to be examined by walking her by the cages of elephants and other primates to gauge her level of fear or curiosity. However, the overall impression upon reading Yerkes's article on the work conducted at the circus, a paper half the length of its predecessors, is one of frustration, compromise, and obstacles to experimental study.

In Search of the Elusive Field

Yerkes's early ventures beyond the laboratory into farms, backyards, and circuses demonstrate the value he saw in studying primates in various novel environments. Such studies were particularly useful when seeking information about primates' basic biology and intellectual capacities. However, the artificial and semi-captive nature of these unusual places for science limited the extent to which primate social behaviors could be explored. To fill this gap in understanding, Yerkes embarked on a program for what he called the "naturalistic" study of primates in the late 1920s. These field studies were planned to supplement work conducted within a laboratory environment and to form one branch of the Yale Laboratories of Primate Biology.[67]

Yerkes developed the program in reaction to concerns about the limitations of the laboratory for the study of certain aspects of animal behavior, while also recognizing the value of studies of animals within their natural social and ecological settings. He viewed such studies as essential for checking the validity of conclusions about primate behavior drawn from the laboratory and for ensuring that the correct methods were used when rearing primates in captivity.[68] Furthermore, fieldwork could provide information about natural primate behavior that was unattainable in a laboratory environment. Specifically, Yerkes identified naturalistic studies as essential for "reliable accounts of the traits, habits, diseases, disabilities, (and) social life, of the wild anthropoid."[69] For Yerkes, other potential places for the study of social behavior, such as the laboratory or zoo, failed to provide such "reliable" information; he even identified his own laboratory studies of primate social behavior as "merely an approximation to fact."[70] Yerkes commented that the results of his laboratory-based study of primate dominance, for example, probably differed from the "situation in nature . . . because of the greater stability of the band and the fact that it is not established by human fiat but by the process of natural adjustment during long association."[71]

For the biologists working at the Yale Laboratories of Primate Biology to gain knowledge of the natural social behavior of primates they had to venture into tropical areas and unknown methodological terrain. The first two naturalistic studies in Yerkes's program demonstrate the methodological obstacles faced by field investigators attempting to observe primates in the wild. In 1929, hot on the heels of Yerkes's study of the gorilla mind in Burbridge's backyard, Harold Bingham, the first fellow of Yale Laboratories of Primate Biology to leave for the field, traveled with his wife to the heart of the Belgium Congo to study gorillas in the wild.[72] Yale sponsored the expedition, and the Carnegie Institution donated $9,000 for the research. It was hoped that the expedition would yield information about the evolution of gorillas in their natural habitat and provide supplementary theoretical information for laboratory studies and practical information concerning the breeding and maintenance of gorillas in captivity.[73] Both Bingham and his wife lacked any field experience in tropical areas and thus drew upon the experiences of others, including explorer Mary Jobe Akeley (the wife of Carl Akeley), when preparing for the field.[74]

The involvement of the Akeleys in Bingham's research went beyond providing methodological advice. The study's location, for instance, was a site the couple had created as part of their efforts to establish an area for the protection and scientific study of gorillas.[75] Carl Akeley began discussing a gorilla reserve in Congo in 1922. Three years later the Albert National Park—named in honor of King Albert of Belgium and now called Virunga National Park—was established, fulfilling Akeley's dream of a sanctuary for gorillas. After his death in 1926, it fell to his widow to steer the American branch of the park's International Scientific Committee. This group included Yerkes and other eminent researchers such as John C. Merriam, president of the Carnegie Institution of Washington, and Henry Fairfield Osborn, president of the American Museum of Natural History. Following Akeley's wishes, the committee encouraged scientific studies of gorillas. Thus, in many ways Bingham's study was connected with Carl Akeley's dream of a gorilla sanctuary that served to increase the world's scientific knowledge of primate behavior.[76]

Unlike the Akeleys, Bingham and his wife had no experience studying wild primates in tropical conditions. In an attempt to gain some basic training, Bingham spent time at Rosalia Abreu's estate in Cuba.[77] As described above, the primates were semi-domesticated and kept in small cages, ensuring very different working conditions from those Bingham would encounter in the Belgium Congo. A further deficiency of Bingham's training was the lack of consultation with Ben Burbridge, who had previously successfully observed gorillas in the wild. This oversight further confirms Yerkes's view of Burbridge as a hunter-collector rather than as a scientific researcher or collaborator.

This poor training, combined with the elusive nature of gorillas, led to great difficulties in observing these primates in the wild. During a brief period of optimism, Bingham commented to Yerkes that trailing the gorillas was "scarcely more difficult than I found it as a boy to follow rabbit tracks in a fresh snow." But shortly thereafter Bingham confided to Raymond Dodge, a psychologist at Yale, that his early success at observing the primates had been little more than "beginner's luck."[78] He vented his frustrations at working in the field to Dodge and recounted his attempt to rehearse methods to observe the wild primates: "The guides led the files, Lucille and I followed, the gun bearer next, and the porters last. The guides located our 'gorillas,' we all retreated for concealment. . . . Well I had to leave my place of concealment to get the men in their proper places, then return to my own station, and watch a hopeless rehearsal. If the act can ever be put over, I predict a downpour just at the moment the gorillas go by."[79] These methodological difficulties resulted in the couple literally turning to the trail for their field observations.

In place of actual observations of gorillas, Bingham and his wife mostly relied on gorilla nests to guide their analysis of primate behavior. Their failure to make field observations was demonstrated in the text of Bingham's published account of the study and reinforced by poor photographs of shadowy, distant gorillas.[80] A crucial ingredient in field observations of gorilla behavior, the presence of the primates themselves, was largely missing. Reflecting on his time in the field three years later, Bingham appeared to hold the African "natives" at least partially accountable for this deficiency: "I could train them on sighting gorillas, but it was probably because they were relying on other senses. However, I think their reputed 'sixth sense' is pure fiction, for there are other denizens of the forest that are tell tales like the blue jays in the New England forests. As to native noses, I believe the boys we had were generally low in olfactory acuity."[81]

The absence of field observations of wild gorillas did not go unnoticed. An unidentified friend of Ada Yerkes saw a presentation Bingham made shortly after his return from the Congo in front of representatives from the Carnegie Institution. The friend wrote to Ada expressing her disappointment in Bingham's results, a feeling apparently shared by many who saw the presentation:

> Perhaps we expected too much in the way of conclusions and observations of gorillas in Africa—about the only things he told were that gorillas

seemed to spend only one night in each "nest," that they sometimes made use of hollow or overhanging tree trunks, or mashed down the tops of trees for nests, that they soiled their nests with feces.... He showed ten or a dozen large clear stereopticon pictures of feces, but only two or three pictures in which we could actually see gorillas, and there the gorillas were so far away we could not in some pictures see them nor distinguish them from leaves till Dr. Bingham pointed them out with his pointer.

Ultimately, this member of the audience concluded that Bingham's introduction, which focused on chimpanzees and was accompanied by photographs of chimps at Franklin Farm and the Yale lab, "was more interesting than anything he told us about Africa—as we could *see* the chimps!"[82]

Although Carnegie's staff approved of Bingham's work, Yerkes came to share the sentiment expressed in the above letter, later refusing to offer Bingham a permanent position in the Yale Laboratories of Primate Biology.[83] This marked the beginning of ongoing emotional strain for Bingham, who had recently lost his father and failed to gain permanent employment at Yale but, despite the difficulties he had encountered in the Congo, yearned to continue field studies of primates. At one point, Bingham was so desperate to find a source of income with which to pursue primate studies, he briefly considered cowriting popular works on life in Africa with Richard Sparks, even though he feared such publications would result in his "fall from scientific grace."[84] Yerkes and Bingham continued a relationship in the form of occasional letters and requests for references. However, Bingham failed to successfully reenter primate studies.[85]

In 1929 Henry Nissen, the second fellow at Yale Laboratories of Primate Biology, traveled to western Africa to observe a less elusive primate, the chimpanzee. Nissen had received his MA and PhD in psychology at Columbia University in New York and would develop into a key figure in both the intellectual productivity and logistical running of the Yale lab.[86] Once again, this expedition was motivated by the laboratory's need for information about certain aspects of primate behavior that could not be adequately studied under captive conditions. Yerkes also entrusted Nissen to forge the beginnings of a cooperative relationship between the Yale Laboratories of Primate Biology and the Pasteur Institute of Kindia. This relationship was part of Yerkes's attempt to "internationalize biological research" to build the necessary connections for naturalistic observations of primates, studies that would inevitably lead American primatologists to Africa and Asia.[87]

Furthermore, Nissen was to return with chimpanzees for captive colonies at both the New Haven and Orange Park Laboratories. This collection of chimpanzees for the Yale Laboratories was the trip's primary objective, with the observation of chimpanzees in the wild being a secondary concern for Nissen.[88] Nevertheless, he spent two and a half months in the field with a total of sixty-four days spent doing "active field work." Chimpanzees were actually observed on forty-nine days.[89]

Nissen, like Bingham, turned to the time-honored techniques of naturalists, such as blinds and trailing, to observe the primates with varying degrees of success. Unfortunately, these techniques failed to enable systematic observations of primate behavior. Although Nissen successfully observed chimpanzees in the wild, his field observations provided sufficient evidence to merely "guess" about the behavior and structure of primate societies, such as the number of primates in a group. It was clear that traditional naturalist practices were insufficient to guide the observation of primates in the wild. This result was of no surprise to Nissen, who had previously agreed with Yerkes that one goal of the expedition was to test "the feasibility of field studies" and to make "a start at discovering workable methodology and techniques for naturalistic observation of the chimpanzee."[90]

Yerkes was optimistic about Nissen's progress in his quest to observe chimpanzees in the wild. In a letter to Nissen, Yerkes wrote: "We think of you now as hard at work in proximity to some native village, following the chimpanzee as a brother and making friends with him."[91] Despite Yerkes's naive impression of what work was like in the field, Nissen experienced frustrations similar to those expressed by Bingham. In an excerpt from a letter written to Yerkes in 1930, Nissen described the difficulties he encountered in both using blinds and in avoiding scaring the primates being observed:

> The animals are extremely timid and astoundingly sensitive to the approach of man.... During these first few weeks in the field I have seen chimpanzees almost every day, but never for more than 45 minutes at a time.... For the last two days I have tried the method of blinds, but without success; the chimpanzees refused to come to the place where the blind was! Please do not think that I am discouraged—far from it. But it is clear that my task is one which requires infinite patience and which yields results only slowly.[92]

Nissen continued to vent his frustrations about the use of blinds and his attempts to observe natural primate behavior to Yale psychologist, Otto L.

Tinklepaugh, complaining that the chimps avoided the blinds and were easily spooked. When he did see the animals, he feared his very presence reduced the naturalness of the primates' behavior and in turn the utility of his observations: "Several times I have seen a few chimpanzees on the ground, but on these occasions they had taken alarm by the time my observation began, so that their behavior, while interesting, certainly was not 'natural.'"[93] Clearly, success was not going to come easily to the field primatologist.

Studies in barns, farms, private estates, and circus quarters yielded useful data concerning primate intelligence and basic biology and, when performed by an established researcher like Yerkes, had the characteristics and status required to be seen as scientific. However, the artificiality of these spaces made it impossible to research natural behaviors including social relationships and communication. Yerkes strongly believed that field studies were essential for a complete understanding of primate behavior and for a clearer understanding of the connection between humans and nonhuman primates. Thus, primate field studies were an essential supplement to those conducted in laboratory environments.

When venturing into the wild, primate researchers were particularly aware of the need to differentiate themselves and their work from the adventurers and travel narratives that had so dominated popular culture during the nineteenth and early twentieth centuries. This process of professionalization required the application of scientific practices and values to fieldwork. Unfortunately, the unique challenges involved in observing primates in the wild ensured that this was not going to be an easy or fast transformation.

The methodological challenges encountered by Bingham and Nissen demonstrate the early stage of development of field methodology for the study of primates. Traditional naturalistic practice was clearly not sufficient for prolonged observation of primate social behavior. Thus, at the close of the 1920s, researchers had attempted but failed to make a systematic, comprehensive, study of natural primate social behavior and, in turn, failed to escape the specter of primate folklore and amateur adventure.

Fortunately, Yerkes's third postdoctoral fellow, Clarence Ray Carpenter would have far more success as he traveled to Barro Colorado between 1931 and 1933 to study howler monkeys. Unlike his predecessors, Carpenter would make significant contributions to the development of field techniques, creating a methodological platform with which to advance primate field studies.

3

Control, Repetition, and Objectivity

Turning Field Observation into a Science

By the end of the 1920s, the first two attempts by scientists to study natural primate behavior had come up short. Harold Bingham's observation of shadowy, distant gorillas and Henry Nissen's preliminary study of chimpanzee behavior had made it clear that researchers lacked both the techniques and tools for observing primates in the wild. Clearly, distancing primate science from the stories of hunter-adventurers was not going to be easy. Thus far, the amateurs had enjoyed far more success observing primate behavior in the wild than had their scientific counterparts, at least if you believed everything they wrote in popular periodicals. Nevertheless, these initial challenges failed to lessen Robert Mearns Yerkes's commitment to developing field studies of primate behavior. Nor did they dampen the spirits of Yerkes's third postdoctoral fellow, Clarence Ray Carpenter, who developed a host of new tools with which to carve out space for field primatology as a real science, separate and distinct from myth and exaggeration.

First, as a postdoc studying howler monkeys in Panama and later as a young assistant professor observing gibbon communication as part of the Asiatic Primate Expedition (APE) in the jungles of Thailand, Carpenter made significant strides in establishing scientific rigor for primate field studies. From determining how to get an accurate count of individuals to interpreting complex social behaviors and recording primate calls, he significantly expanded the methodological arsenal for field primatology. Unlike his predecessors, Carpenter managed to find and systematically study primates in the wild and develop new methods for observing and interpreting primate social behavior. Much of his time in both Panama and the Pacific was spent developing ways to break down complex social behaviors into readily observable and replicable parts. His census and recording techniques allowed for the first accurate counts of wild primate populations and for repeated observation of their forms of locomotion, social behavior, and vocalization. Together, these methods and tools demonstrated that field studies of primate behavior could fulfill the values laboratory scientists espoused: repetition, accuracy and, in the case of vocalizations, mechanical objectivity.

Carpenter's development of new methods allowed for theoretical insights that never would have been possible without seeing primates interacting in the wild. Carpenter's study of howlers, and his later research on gibbon vocalizations, demonstrated that primate calls were far more complex than first thought. Instead, the calls of howlers and gibbons served vital roles in maintaining the social cohesion of groups. The preliminary conclusions formed during his time in Panama were confirmed with the aid of recording technology, which he used during the APE to enable repeated observations of gibbons' responses to calls.

Carpenter's fieldwork in Panama and then Thailand also highlights a transition occurring in primate studies during the interwar period, from field researchers working in relative isolation to being a member of a clearly defined team. Prior to the APE, Carpenter was an individual Western researcher, at times accompanied by his wife, who conducted field studies with the aid of local people serving as porters or guides. In contrast, the APE was a team-based effort that involved several American colleagues and the assistance of a dozen or more Thai porters who carried various camp and science equipment into the jungles of Thailand.

Third Time's the Charm:
Studying Howler Monkeys on Barro Colorado

Born in 1905 in Lincoln County, North Carolina, Carpenter received his master's at Duke University in 1929 working with William McDougall on a study of sex behavior in pigeons, specifically the monogamic tendencies of male pairs. Carpenter continued researching pigeons for his PhD, which he gained in 1931 under the direction of Calvin P. Stone at Stanford for his work on the social behavior of birds. At the time, Stone was using funding from the Committee for Research on Problems of Sex (CRPS), the National Research Council (NRC) subcommittee chaired by Yerkes, to support his graduate students. Thus, it is not surprising that Carpenter's doctoral research focused on the connection between physiology and sexual behavior.

Carpenter's doctoral research involved a laboratory study of the impact of castration on male sex drive. He used still and motion pictures to confirm his observations, allowing him to see behaviors that required multiple viewings to identify. The publications that resulted from this doctoral work, in which he critiqued the effectiveness of current surgical methods for gonadectomy and developed an observational system for studying sexual behaviors in pigeons, reveal Carpenter's interest in methodological issues.[1] After dismissing several potential methods of observation on either practical grounds or out of concerns that they would modify the birds' sex drive, Carpenter settled on the "time-limited systematic observational technique." This was a modified version of a method that Stone, his doctoral advisor, had used to study the sexual behavior of albino rats. This method appealed to Carpenter because it allowed for the collection of quantifiable and consistent data.[2] Observing the birds for exactly ten minutes twice a day and noting their positions and all sexual activity on "standard record sheets" achieved these goals.

To maximize the expression of sexual behaviors, Carpenter used two different types of experimental frameworks. In "A-test" observations, he separated male and female birds for twenty-three hours and then observed all sexual behaviors between the newly reunited birds for a period of fifty minutes. In contrast, "B-test" observations involved recording sexual behaviors between birds without a prior separation period. Throughout the study, Carpenter used motion pictures to record comparisons between normal and castrated birds. The use of film provided several advantages, including the possibility of accurately timing interactions and creating a record of behaviors that were difficult to capture by verbal description alone.[3]

Carpenter's interest in combining observation and experimentation to yield accurate and, when possible, quantifiable data increased as his career steered away from laboratory studies of birds and toward field studies of primates. Entering an area desperately in need of techniques to see and understand wild primate behavior, Carpenter repeatedly impressed established researchers like Yerkes with his skills for observing and recording animal behavior. Carpenter also took pains to emphasize the ability of fieldwork to meet scientific standards commonly associated with the laboratory, differentiating his work from his amateur predecessors on the grounds of accuracy, repeatability, objectivity, and quantifiability.

For his masters and doctoral work, Carpenter focused on the development of methods for the study of sex. His interest in sexual behavior also characterized his primate research that, like his bird studies, was at least partially funded by the NRC-CRPS.[4] Animals allowed for the kind of controlled, experimental, and quantifiable study of sexuality that was believed to be impossible for human subjects. Thus, in the 1920s and 1930s it was common for animals ranging from birds to primates to be used as analogies for human sex drive, copulation, masturbation, and homosexuality. The sole focus of his laboratory bird research, these sexual topics inspired some of the dozens of questions pursued in Carpenter's primate field studies, including how primates moved, operated in groups, maintained territory, and communicated, as well as exploring primates' estrus cycle and mating behaviors.

Carpenter's first sustained field study of primates began in 1931 with his research on the social behavior of howler monkeys on Barro Colorado, Panama, which Carpenter conducted while he was a postdoctoral fellow working with Yerkes at Yale. He completed this research in 1933 after spending a total of nearly eight months in the field. In between trips to Barro Colorado, Carpenter married his first wife, Mariana Evans Carpenter, who at times accompanied him on field trips.

Barro Colorado bestowed several advantages. An island in Panama formed as a result of artificial dams, it was established as a wildlife reserve in 1923 and was maintained by the Institute for Research in Tropical America (IRTA). The IRTA was created in 1921 as part of the NRC's attempts to establish cooperative research on the tropics and included representatives from the American Museum of Natural History, the American Society of Agronomy, and the National Geographic Society, to name but a few of the organizations that supported it. Despite the involvement of several powerful stakeholders in the IRTA, the body failed to continue beyond the 1920s and the island came under the control of the Smithsonian Institution.[5] Although covered in "wild" forests, with indigenous flora and fauna, the island was shaped to maximize the ease with which scientists could observe animal behavior. Researchers and animals used human-made trails, for example, allowing scientists to see animals away from the dense foliage of the tropical jungle.[6]

The ornithologist Frank M. Chapman described Barro Colorado as an "island laboratory" that benefited from being isolated by water and rich in indigenous flora and fauna.[7] In Chapman's words: "Living conditions at the laboratory so nearly approach the luxurious that one constantly feels in an apologetic frame of mind toward those naturalists who, to reach primeval surroundings, have gone farther, fared worse, and seen less."[8] Indeed, Barro Colorado offered a seemingly perfect combination of safety, comfort, and productivity for biological research. Chapman was instrumental in initiating primatological research on the island, writing to Yerkes to inform him that howlers could be observed with relative ease on Barro Colorado. This letter spurred Yerkes to seek a researcher for fieldwork on the island, a job for which Carpenter proved particularly well suited.[9]

Barro Colorado's natural physical features and human-made conveniences, such as a laboratory and trails, created an environment that facilitated making prolonged observations of primate groups. Carpenter used this opportunity to develop methods for observing primates in the field and to make the first systematic study of what Alfred Russel Wallace had called "the most remarkable American monkey," the howler.[10] Best known for their "howl," described in earlier works as "a barking roar, drum-like, and like the squeaking of an American wagon" and "as powerful, melancholy, insufferable, and indescribable," howlers were also believed to react negatively to the presence of an observer.[11] Armed with this knowledge and a bag of equipment that included field glasses, notebook, camera, snake kit, and machete, Carpenter made extensive observations of the howler monkey's locomotion, territoriality, and the various social relationships among males, females, juveniles, and infants that characterized primate society. Carpenter also conducted the first census of a wild howler population.

Carpenter's successful observation of the number and behaviors of the howlers on Barro Colorado were possible both because of the nature of the environment and his ability to develop existing and new methods for field science. For example, Carpenter developed a technique for precisely counting the number of individuals in a primate group. Through observing howler monkeys, he realized that they moved in single file when progressing through trees. Rather than attempting to count and identify individuals when they were congregated together, he counted and identified the sex and age of the individuals during their progression through trees. This approach, which grew out of ornithology, greatly eased the challenge of conducting a census of wild primate populations.[12] Nevertheless, completing a count of a howler group often took several hours, and at times days, to complete. Although still time consuming, Carpenter's counting procedure was certainly a marked improvement on Nissen's attempt to count primates by encircling the group in a failed attempt to entrap them. In contrast to Nissen's approach, which did not yield an accurate count, Carpenter's census determined there were 398 howlers on Barro Colorado in 1932, plus or minus fifty as the "probable error." The following year, the census was repeated with a count of 498 howlers, with an increase from twenty-three to twenty-eight primate groups.

In addition to developing a technique for counting primates in the field, Carpenter established practices for observing and recording the complex social interactions that characterized primate groups. He did this by first identifying the age and sex of each individual. Adult males, who were larger and hairier than females and frequently the ones to lead the group and react most vocally to the observer, were relatively easy to identify. Females were identified as either carrying or closely associated with young. Juveniles posed the biggest challenge when it came to identification, because juvenile males could easily be mistaken for females. Aware of this risk, Carpenter made "every possible check ... to avoid a systematic error of this kind."[13]

Determining the age and sex of each individual allowed each primate group or clan to be broken down into the social parts that formed the whole. Under the heading female-young relationships, for instance, Carpenter included behaviors such as maternal aid, retrieving fallen young, and weaning. Summaries were supplemented with specific numbered observations, serving as examples of the types of behaviors observed and the types of notes recorded. This approach, which became known as the dyadic method, was repeated for male-female relationships, male-male relationships, and so forth. Observation 59, included in the male-male relationship section, exemplifies the style of Carpenter's observations:

> Feb. 21, 1932, Group 1. I was in an arboreal blind at Barbour 2 observing Group 1 when males began howling rapidly and ferociously. For two hours and a half the intermittent roars continued.... From my location I could not discover a provocative cause and I left my blind to search more carefully. Fifty yards away toward Shannon 1, I found a complemental [prospective new or temporary group member] male. This animal was a fine specimen and approached within 25 feet of me, behaved somewhat aggressively, and then yawned. With continued roars coming from the clan, the complemental male withdrew in the direction of Shannon Trail.[14]

As in Carpenter's doctoral work, these types of observations were confirmed with the use of still and motion photography with the goal of increasing their accuracy and objectivity.

At the heart of many of Carpenter's methodologies was a desire to scientifically observe what he termed *natural behavior*. Carpenter tried to ensure this by using three strategies to avoid disturbing the primates. Each observational technique included checking and rechecking in order to ensure accuracy. The first observational technique used to prevent the disturbance of the primates was the use of blinds, a technique that originated within the naturalist tradition and was commonly used in ornithology during the 1920s.[15] The second method, again of naturalist origin, was the use of various landmarks, such as embankments, from which to view primates while remaining hidden from their sight. Of course, Carpenter's attempts at hiding were not always successful, and howlers who spotted him reacted by roaring, hiding, or defecating.

Carpenter's third method, and the technique that had the greatest impact on field primatology, was the establishment of "tolerance distances" from which to observe howlers without undue modification to their behavior, a process now known as *habituation*. When observing a clan of howlers identified as Group 1, Carpenter sat so as to be visible to them. Gradually Group 1 became "neutrally conditioned" or habituated so that by 1932 Carpenter could approach them and "feeding, play, or progression continued and no defensive roars were given."[16] Such habituation was essential in an environment in which the usefulness of traditional naturalist methods, such as blinds, was limited due to primates only occasionally taking the same routes through the jungle. Therefore, Carpenter supplemented these traditional methods with his habituation technique.[17]

Despite Carpenter's concern about preserving natural behavior, he also used experimentation, albeit to a limited extent, while studying howlers on Barro Colorado. For example, when attempting to determine whether howlers could swim, Carpenter removed an adult female to a nearby island. Her later return to the main island of Barro Colorado led Carpenter to conclude she had swum at least fifty feet, although he did not observe the animal in the water. He also made mention of two "natives," both "dependable" and "unusually keen" observers, who saw adult male howlers swimming " 'mismo un hombre'—like a man."[18]

Interpreting Primate Voices: Studying Primate Communication on Barro Colorado

When embarking on his study of howlers, Carpenter was aware that animal calls were commonly believed to serve little, if any, social role. In his words: "It [primate vocalization] has been supposed to serve no useful function whatever, to correspond to the song of birds and be a kind of music, to be an expression of emotions, and to serve the function of defense."[19] During his time on Barro Colorado, Carpenter became convinced of the importance of vocal cues for social behavior. He came to identify vocalizations as functioning to stimulate group progression and to locate lost or distressed juvenile group members. A "deep, hoarse cluck" sound made by a leading male howler, for example, stimulated the clan to move, a behavior called *group progression*. Carpenter observed "hundreds of instances of the sequence of vocalization and behavior" and became "convinced that the deep cluck [was] given as a signal which initiates and directs, as its absence inhibits, group progression."[20] Similarly, the "wailing groan of the mother" served to encourage a lost infant to move toward her, while also provoking male group members to search for the young howler.[21] Vocalizations served as "facilitators," stimulating individual and/or group action, while other calls functioned as 'inhibitors,' preventing actions such as juveniles moving too close to the observer.

Carpenter's understanding of primate communication as essential for social coordination and control represents both a continuation of and contrast with earlier works by Yerkes and Wolfgang Köhler, a psychologist from Germany who studied primates at the Anthropoid Station in Tenerife and published *The Mentality of Apes* in 1925.[22] In 1929, Robert and Ada Yerkes's seminal work, *The Great Apes*, stated that chimpanzees understood "not only the expression of *subjective moods* and emotional states, but also of definite desire and urges."[23] Although Yerkes at times flirted with the idea of primates having the capacity for intercommunication beyond emotional expression, he ultimately stayed within an affect-based understanding of primate communication. Although Köhler did not fully accept this established view, he failed to offer an alternative theory of primate communication such as that later provided by Carpenter.[24]

Yerkes's views of primate communication also varied somewhat depending on the species of primate. For Yerkes, the chimpanzee demonstrated the highest level of intelligence within the nonhuman primate order. Thus, he identified chimpanzees as having the highest ability for intercommunication both with each other and with humans. Gorillas and orangutans demonstrated some ability for intercommunication; however, he saw no evidence for such ability with gibbons.

Yerkes's use of captive subjects with his laboratory studies restricted his understanding of primate communication to the calls of one or two individuals. Thus, he rarely observed what he termed "intercommunication" between primates, and when such communication was recognized, it was impossible for him to locate it within a truly social context. Carpenter, in contrast, observed the role of vocalizations in the coordination and control of a complex society within its natural ecological setting. In the laboratory, Yerkes observed communication between only a few primates, and such communication was restricted by the artificial nature of the primate's environment, including the lack of social groups, adequate space within which to range, and the absence of other species that would have been present in their natural habitat. Thus, Yerkes did not discuss the social function of vocalizations, a central concept in Carpenter's understanding of primate communication. This problem was amplified by Yerkes's bias toward the study of primate vocalizations in terms of their ability to imitate human language. Yerkes's goal was not to gain an understanding of how communication functioned within primates' natural social and environmental context. Rather, he sought an understanding of the connections between humans and their cousins, of which communication represented one important thread.[25]

When developing his understanding of primate communication, Carpenter looked to studies of other species, including birds. In 1969, Carpenter reflected back on the sources upon which he based his knowledge of the function of vocalizations when first entering the field in 1931.[26] He recalled that among his main influences was ornithologist Wallace Craig's 1908 paper concerning the use of vocalizations to synchronize and coordinate behavior.[27] Craig argued that social instincts alone were insufficient to enable individuals to succeed in their changing social environment. Instincts needed to be coordinated with the actions of other members of the group. The voice, he argued, served this function. In Craig's own words, the voice was a "social influence" that brought "social instincts" into "harmonious co-operation."[28] Carpenter echoed this interpretation with an emphasis on the voice as a means of social control and group integration.[29]

Henry Elliot Howard was another ornithologist who influenced the development of Carpenter's views of primate communication.[30] Howard's study of bird behavior played a significant role in the evolution of the concept of territoriality, specifically the role of birdsong in the maintenance of territory. This was one of the few functions of vocalizations adopted by Carpenter prior to his study of howler monkeys in Barro Colorado. He would later connect this function with the avoidance of conflict and thus the reduction of aggressive behavior. He would eventually expand the social function of vocalizations to include a range of social coordination and control mechanisms.[31]

Beyond ornithology, behaviorism shaped Carpenter's understanding of animal communication. In particular, Grace de Laguna, an advocate of a blend of behaviorism and an experimental psychology approach called Chicago school functionalism, who published the 1927 book, *Speech: Its Function and Development*, significantly influenced Carpenter's thinking about primate communication.[32] Through de Laguna, Carpenter learned the importance of observing "the responses of the associated animals" when studying the function of animal vocalizations.[33] Furthermore, for Carpenter, it was de Laguna who placed *social* behavior at the center of a functional understanding of animal calls. De Laguna argued that animals had to be studied within the context of a social group. Previous studies had failed to conceptualize animals as individuals within organized societies and thus failed to investigate the social function of communication. Craig, too, referred to this dearth in studies of social communication in his 1908 paper, writing: "The inadequacy consists in studying the birds as individuals, and in treating the individual as a distinct entity. What is needed is to transcend this individualistic view-point, and to see that the instincts of the individual can affect their purposes only when they are guided and regulated by influences from other individuals."[34] The voice of the pigeon was one such influence.[35]

Grace de Laguna was one of an emerging group of scientists who challenged the traditional view of animal communication that animal calls merely served to release emotional pressure within individuals. In *Speech: Its Function and Development*, she sought a new approach to animal communication, one based upon the central concept that animal cries served a social function. She complained that, as of 1927, this theoretical transition had failed to be incorporated into experimental studies of animal communication. Although she saw an "interesting beginning" in the work of Yerkes and Köhler, de Laguna argued that both had failed to analyze the function of animal vocalizations systematically.[36] However, in Craig's 1908 article de Laguna had found a "worthy exception" to the rule.[37] Craig's views on pigeon communication exemplified her thesis that animal cries served an important social function within animal communities. De Laguna also commented that there remained a strong need to expand this interpretation of animal calls to the study of "gregarious mammals." Carpenter ultimately provided such a study.[38]

Concerning the issue of whether primate calls served a significant social or purely affective function, Carpenter clearly sided with those who argued that vocalizations played a central role in primate societies. He understood primate communication as expressing emotions such as fear and excitement but also as serving to coordinate and integrate primate groups. Influenced by Yerkes, Köhler, Craig, Howard, and de Laguna, Carpenter was able to offer a new, truly social, understanding of primate vocalizations. This theory was well formed after his field studies of howler monkeys on Barro Colorado in 1931–1933 and was verified four years later during the Asiatic Primate Expedition with the aid of recording technology.

Technology and Repetition in the Field: The Asiatic Primate Expedition of 1937

In 1936, Harold J. Coolidge, a naturalist who worked at the Museum of Comparative Zoology at Harvard University, began organizing a six-month foray that would become known as the Asiatic Primate Expedition. He immediately thought of Carpenter as a potential collaborator. Writing to Carpenter in February of that year, Coolidge made his pitch: "To my mind there is no one living who is better qualified to make such a survey as I have in mind than yourself." Coolidge had in mind an expedition focused on the gibbon, with Coolidge also planning to conduct a study of orangutans in North Sumatra in cooperation with the Netherlands Indian Society for Nature Preservation. He envisioned a group of American researchers divided into specializations with some focused on collecting dead specimens while others observed the locomotion, vocalizations, and social behaviors of the living. Coolidge was responsible for acquiring funds for the expedition and leveraging his wide network of professional contacts to facilitate logistic planning in both America and the Far East. Yerkes played a central role in these discussions, consulting both directly with Coolidge and with Carpenter concerning the organization of what they called "a cooperative gibbon program." Carpenter would not agree to be part of the expedition without first discussing the matter with Yerkes, his former postdoctoral advisor and mentor.[39]

Coolidge's courting of Carpenter was successful, with the main expedition team ultimately comprising of Coolidge, Adolph H. Schultz of the Department of Anatomy at Johns Hopkins University, and Carpenter, who at that time was an assistant professor and lecturer at Bard College. Sherwood Washburn, a graduate student of Earnest Hooton at Harvard who would go on to become a prominent anthropologist, was also part of the team, as was Augustus Griswold Jr. who collected birds and small mammals. John T. Coolidge Jr. served as an artist and photographer. Other collaborators included H. G. Deignan, an ornithologist working for the U.S. National Museum in Siam, and an unidentified Thai assistant. Although she is not mentioned in Carpenter's publications or in any detail in his field notes, apparently Carpenter's wife, Mariana, came along, too, as he listed her among the "Personnel of [the] Asiatic Primate Expedition" in a record at the back of his APE field notebook.[40]

Gibbons' occasional bipedalism and apparent monogamous family structure led some scientists in the 1930s to identify them as evolutionarily closer to humans than we commonly do today. Indeed, in his introduction to Carpenter's 1940 APE monograph, Coolidge described them as a member of "that small and select group, the so-called higher primates, consisting of man, the great apes and the gibbons." Gibbons' close connection to humans was also the focus of a 1936 *New York Times* article on the APE that announced, "Scientists to live with jungle apes . . . to trace more of the 'missing links' to man." The article declared that after nine months "every phase of the gibbon's existence will be inspected in light of modern psychology, sociology, morphology, physical anthropology and anatomy in an effort to determine for the first time with scientific accuracy to what extent these primates have followed the same routes of evolution as mankind."[41]

In keeping with the long history of comparing the female genitalia of humans to that of nonhuman primates, Coolidge believed "that the gibbon is much nearer to man than is any other primate regarding the female external genitalia."[42] For Coolidge, it was anatomy, rather than temporary bipedalism, that revealed the relatively close evolutionary relationship between humans and gibbons. His interest in the gibbon was further amplified by his fear that they would soon become a protected species, thus inhibiting his ability to conduct the kind of intensive study of gibbon biology he envisioned. As Coolidge explained in a letter to Yerkes on June 6, 1936, it was this concern that led him to turn his attention away from African apes and toward Asia: "When this happens [gibbons being added to the list of protected species], it will be too late to obtain a sufficient series of one species to make much needed studies of variation. It will also be extremely difficult to obtain reproductive material and expecially [sic] fetuses, which are at present extremely uncommon in the collections."[43]

The plan was for Coolidge to be in charge of collecting dead specimens while Carpenter would be in charge of the "Natural Behavioral Division" of the expedition. However, Coolidge contracted a staphylococcus infection, rendering him unable to do fieldwork. Thus, Schultz and Washburn undertook the majority of the "measuring, skinning, dissecting and preserving" of primate specimens, which they located with the aid of indigenous assistants identified in an informal report as "former boys" employed by filmmaker Martin Johnson. Carpenter meanwhile took over the survey of orangutans in the Sumatra in addition to his study of gibbon behavior. On October 15, 1937, Coolidge summarized the accomplishments of the expedition as having collected "some four hundred documented specimens of primates; about thirty-five hundred birds and mammals from Siam and North Borneo[;] . . . about ten thousand feet of film, much of which is of value in behavior studies on wild gibbons[; and] sound recordings of wild gibbon calls." Ultimately, Coolidge believed "the wild gibbon, instead of being the least known of the four anthropoid apes, should as a result be a strong candidate for the first place."[44]

Although relations between the investigators were not always positive, with Schultz having a negative view of Carpenter's heuristic approach to primate behavior, Coolidge believed Carpenter deserved much of the credit for the APE's success.[45] In a letter to Yerkes, Coolidge particularly highlighted the methods adopted by Carpenter: "I really feel that the amount of information which Ray [Carpenter] gathered on the gibbon was quite astonishing, and I know that it was a revelation to Dr. Schultz. If it had not been for Ray's study we would only have our scant field notes on the behavior of wild gibbons, and these tell very little because they are not tied together by continuous observations."[46] Once again, Carpenter had demonstrated his talent for observing primates in the field.

One notable difference between Carpenter's experience on the APE and his fieldwork on Barro Colorado was the team nature of the expedition. Carpenter was one of an all male team toughing it out in the jungle. Sleeping on foldout beds under a wooden shelter and enduring ticks and mosquitoes, the team embodied pioneering masculinity and fraternity.[47] This ability to survive in the jungle led to Carpenter serving as technical consultant for

Figure 13. Photograph of unidentified porters who worked as part of the Asiatic Primate Expedition. (From C. Ray Carpenter Papers, Pennsylvania State University Archives, Pennsylvania State University Libraries, box 60, folder 1)

a 1944 Army Air Force film entitled *Land and Live in the Jungle*. The video instructed soldiers to conceptualize the jungle as a problem to be solved. When thirsty, search for water in the clear sap of vines, and when hungry observe monkeys as anything they eat "is good for man." If the soldier desired fine dining, they were instructed to make a baking oven, cook a tortoise or bird for two hours, and then enjoy "a dish that would cost five dollars at Wardorf." Certainly Carpenter's ability to not just survive but thrive in the field was an essential ingredient in his success on the APE.

Relatively comfortable beds and access to equipment for examining the bodies of primates and recording their vocalizations enriched the field experience. Getting these items into the field involved hiring local people (see figure 13). However, the expedition's interaction with indigenous people extended beyond transporting goods into the jungle. Whereas mention of locals as sources of information or materials are few in Carpenter's discussions of his work on Barro Colorado, he worked with several local people during the APE who, although unidentified in most of Carpenter's published and unpublished accounts, are acknowledged in his 1940 gibbon monograph. After mentioning those who provided funds and material, Carpenter thanks "those natives who labored to establish and maintain camps under primitive conditions." He went on to say, "I personally have a warm feeling of gratitude for my ancient Karen [a local tribe] guide with whom I communicated only by gestures and for my old Siamese cook whose only attempt to vary the simple camp ration was his prideful preparation of egg custards. I am thankful to the Buddhist priests who allowed me to live in the guest houses of their temples and who, though not understanding my strange activities, tolerated the intrusion on their routine and peaceful habits."[48] Clearly, Carpenter's experience in the field was interwoven with the knowledge, hospitality, and labor of local people.

In both Panama and Thailand, Carpenter used the environment to his advantage. Upon his arrival in Siam, as Thailand was then known, Carpenter had difficulty locating gibbons at first due to dense trees, and thus experienced "a sinking feeling of hopelessness." Soon thereafter, however, Carpenter relocated to a temple village in Doi Dao where, as he had in Barro Colorado, he took advantage of naturally occurring blinds.[49] As he had in Panama, Carpenter made observations "from craggy cliffsides, from hill tops, tall trees or across clearings" and built blinds "at strategic points in outcropping rocks, in bamboo clumps, by springs and near food trees." Doi Dao also had trails that assisted Carpenter's observations, only in Doi Dao rather than being named in honor of prominent naturalists they were made by indigenous peoples "in their search for bark, fruits, roots, orchids and choice teak trees" and by elephants used in the logging trade. As in Barro Colorado, gibbons were concentrated in a small area. The trees themselves seemed to cooperate, soon shedding their leaves, providing Carpenter with a plain view of the arboreal primates. Furthermore, gibbons were protected in Doi Dao, with shooting being prohibited and local people viewing gibbons as "quasi-sacred." Combined, these characteristics led Carpenter to proclaim the site "to be almost ideal for observations of gibbons."[50]

The kinds of tasks performed on the APE were also similar to those in Barro Colorado. Once again, Carpenter focused on identifying the ages and sexes of the primates, performing group counts, examining locomotion, territoriality, and intragroup relationships with the aid of the dyadic method. Like his howler monograph, his gibbon study emphasized the extra measures taken to ensure the accuracy of the counts and the reliability of the observations. Carpenter, for example, reminded the reader that "the reliability and completeness of the first observation of any activity can always be increased by repeated checking."[51] As he did in Panama, Carpenter used still and motion photography as part of this checking process. He also added a new weapon to his methodological arsenal in the form of sound recording equipment that he used to extend his interest in primate vocalizations fostered during his time on Barro Colorado.

Carpenter's use of recording equipment for the study of gibbon vocalizations sits at the intersection of his methodological interests and the development of his theory of primate social communication. Encouraged by Coolidge, Carpenter, or more specifically the numerous indigenous people hired as porters for the expedition, carried a portable Presto recording device with playback attachment, a semi-portable parabolic reflector six feet in diameter, cables, and batteries into the jungle of Thailand with the goal of "study[ing] more completely and objectively the vocalizations of gibbons."[52] The technique was not entirely new; recording technologies had regularly been used to record and stimulate animal calls through the use of playback in both field and laboratory studies of animal communication since the late nineteenth century. In 1937, for example, Arthur Allen, an ornithologist at Cornell, used recording technology and the technique of playback during his mission to record the calls of vanishing bird species. One year later, Ludwig Koch, who specialized in wildlife sounds, and Julian Huxley, the well-known ethologist and evolutionary biologist, required a van to transport the vast amount of recording technology used in their study of the vocalizations of captive animals.[53]

In comparison with the technology used in previous studies by other ethologists, Carpenter's equipment during the APE was smaller and easier to transport. Nevertheless, the wildness of the field site ensured it would remain a challenging task. All told, Carpenter's recording gear weighed at least 170 pounds, despite Coolidge having previously hired members of Harvard's physics department to reduce the weight of the devices.[54] It took a bus and then two ponies and fifteen men walking for one and a half days to get the equipment into the Thai jungle. Once there, Carpenter protected the equipment from the elements, at one point commenting in his field notes that he had to rush to cover the $800 worth of technology during a rainstorm.[55]

Three months into the expedition, Carpenter used their recording technology to successfully record and play back the calls of a single gibbon (see figure 14). The gibbon repeated the calls back to the recording device. Carpenter's discussion of this experience in both his published accounts and

Figure 14. Recording technology used during the Asiatic Primate Expedition. (From C. Ray Carpenter Papers, Pennsylvania State University Archives, Pennsylvania University Libraries, box 60, folder 3)

APE field notebook highlight how playback served to confirm both the behavioral response of the gibbon and the quality of the recording itself: "The time relations and the similarity of quality of calls were strong evidence that a stimulus-response relationship existed between some of the calls as recorded and the vocalizations of the gibbon. There is no better validation of the fidelity of the recording and reproduction."[56] After this success, Carpenter often spent whole mornings and at times entire days recording gibbons' calls, ultimately making at least forty-two recordings.

Coolidge proudly included the use of playback in an "APE Report of Progress" dated March 10, 1937, and Carpenter himself describes the procedure in both a *Scientific Monthly* article in 1939 and in his 1940 gibbon monograph, based on his APE research.[57] The use of this technology formed part of what would become Carpenter's career-long endeavor to apply rigorous scientific standards to animal field studies, which he felt had a "retarded reputation," through the application of scientific standards to fieldwork.[58]

Reflecting back on the APE, he commented: "An attempt was made in 1937 while studying the vocalizations of gibbons in Northwestern Thailand to advance the methods, equipment, and technologies of this kind of research."[59] Here, Carpenter was referring specifically to the recording technology used during the APE and its application for both the recording of samples of gibbon vocalizations and the use of playback to verify their functions.

Carpenter's views on the important social functions of primate communication were already well formed by the time he embarked upon the APE. Thus, the use of playback experiments served primarily to confirm, rather than develop, Carpenter's function-based understanding of primate communication. Carpenter would note which responses would be stimulated by certain calls and thus determine their function. The purpose of the vast majority of the calls was social and, therefore, revealed the importance of communication in primate groups. In Carpenter's own words: "The playback capabilities of the relatively primitive recording-reproducing equipment permitted me to reflect back the recorded calls to the group of gibbons which had just made them. Thus, a procedure was originated in miniature for validating calls and for checking their fidelity in terms of responses as well as providing a means of controlled study of the functions of different sound signals."[60]

Carpenter recognized that the use of technology in the field for recording primate vocalizations was not without limitations:

> The recording of vocalizations of gibbons revealed some of the potentials as well as the limitations of this procedure. Surely field studies of vocal behavior are needed to complement laboratory studies and to ascertain the functions and operational characteristics of vocal signals. However, field recordings are as yet difficult to make without disturbing the wild subjects. Also, the recordings of DISTANT CALLS are easier to make than the CLOSE in-group calls; hence, the samplings of the full repertoire of a species is subject to bias favoring the louder and distant calls.[61]

Here Carpenter expresses an argument he, along with other animal researchers such as Yerkes, commonly repeated: complete knowledge of primate behavior required the combination of field and laboratory studies. In the case of vocalizations, field studies should be used to determine the function of calls due to the limitations of the lab for the study of natural behavior. Lab studies, in contrast, were well suited for the study of quiet intragroup vocalizations and for the isolation of calls from surrounding sound stimuli.[62]

Despite these limitations, the use of playback technology had clearly demonstrated that technology could be used to enable field observations to be repeated and interpretations confirmed. This ability, combined with Carpenter's methods for counting individuals, breaking down complex social behaviors, and maintaining the naturalness of the behaviors observed led others to identify his work as the first truly scientific study of wild primate behavior.

Through the development of field practices and the application of technology during the APE, Carpenter sought to demonstrate that scientific standards could indeed be realized in the field. In Carpenter's own words, work done in the field was "as objective and as scientific as the ingenuity of the worker permitted."[63] In Carpenter's work, this is particularly well demonstrated by the use of recording technology during the APE to record and play back primate vocalizations in order to confirm his theory of primate communication. Such technology allowed behaviors to be repeatedly observed, therefore, increasing the scientific credibility of field observation.

For these reasons, Carpenter's field methods and sites served to exemplify how fieldwork could meet scientific criteria. His studies introduced an emerging generation of primatologists to fundamentals of primate observation, opening the eyes of the scientific community to the potential of primate studies.[64] Even beyond the realm of academic and popular science, Carpenter was shown as an innovator in field science in a 1973 children's book teaching young people how to be good observers and naturalists by following in the footsteps of his fieldwork in Barro Colorado.[65]

Although in the eyes of others, Carpenter's fieldwork at Barro Colorado and during the Asiatic Primate Expedition were obvious shining examples of what fieldwork could, and should, be, Carpenter's own perception of his career and the field as a space for science was more complex and contradictory. After returning from the APE, he became a vocal defender and promoter of field studies. However, underlying his post-APE research were concerns about the field's ability to offer efficient means of data collection. In response to these nagging doubts, Carpenter created alternative spaces for fieldwork while also creating dynamic definitions of what constituted natural primate behavior. It is to this interesting and complex quagmire of contradictions to which we now turn.

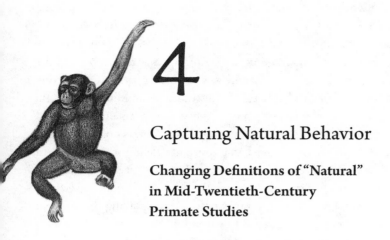

4

Capturing Natural Behavior

Changing Definitions of "Natural" in Mid-Twentieth-Century Primate Studies

In the 1920s, Robert Mearns Yerkes was telling anyone who would listen that the field was distinct from the lab because the behavior observed there was "natural." A decade later, Clarence Ray Carpenter took up the mantle of advocating for field studies of primates. Like Yerkes, Carpenter held up naturalness as the source of the field's uniqueness. At the same time, Carpenter increasingly applied laboratory criteria to his fieldwork. Carpenter and others identified his field studies of primates as embodiments of good quality science because they were accurate, objective, and supported by repeated observations, both in nature and through the replaying of still and motion pictures and sound recordings. His field studies were accepted as scientific because they were, in many ways, like those in a lab.

Early primatologists wanted to have their cake and eat it too. They understood field studies of primate behavior as special because natural primate behavior could only be observed in the wild, but they also wanted primate field studies to be seen as scientific, which for them meant making the field like the laboratory in several key ways. Both lab and field studies used experiment, observation, repetition, and quantification whenever possible; both accepted without question that these were the criteria of good science.

Rightly or wrongly, field primatologists, similar to field researchers in other biological sciences, felt that others saw them as "second-class" scientists, to use the words of geneticist Francis Sumner.[1] Many of them believed that the field had what Carpenter referred to as a "retarded relative status" in comparison to the laboratory.[2] In the 1930s and 1940s, Carpenter and his like-minded colleagues embarked on a two-pronged attack to bolster the reputation of the field and the researchers who worked there. On the one hand, they wanted to demonstrate that field studies could fulfill the criteria they articulated for "good science"; on the other, they proclaimed the field as special on the grounds of the naturalness of the behaviors observed there.

The concept of specialness played an important role in the development of field studies during the 1930s and 1940s. Historian of science Robert Kohler, in particular, has traced the ways that field biologists "assert[ed] specialness to attract talent" in the 1920s before downplaying the concept a decade later.[3] Primatology followed this trend in some ways but not others. Unlike older field sciences like ecology that had more of a foundation on which to build credibility for their work, primatology was still a fledging science in the 1930s, and it existed under the long shadow cast by centuries of primate folklore. More than any other young field science, primatology had to find methods and terms with which to claim authority and credibility. The scientific community and the public had to identify primatology as legitimate science, distinct from past amateurs' tales of beasts in the jungle. At the same time, early primate researchers continued to believe their scientific subjects provided unique gateways to understanding human aggression, sexuality, and social relationships. They also believed that only in the field could natural primate behavior be observed. Laboratory studies could only offer partial answers to questions about human behavior. It would be necessary to turn to nature if we were to ever understand our natural selves.

This challenge of declaring their work as simultaneously special and similar defined the work of emerging primate researchers. For epistemological and practical reasons, picking up the methods, tools, and terms of the laboratory and simply transplanting them into the field was not going to work. Studying primates in the wild is vastly different than observing them in the laboratory. Carpenter, the trailblazer of field primatology, did what many field naturalists were doing in the 1930s. He took methods originating in natural history and injected into them "some of the analytical force of laboratory precision and causal analysis."[4] Carpenter's early field studies were thus of the field at least as much, or more, as they were of the lab.

Combining lab standards with field methods and values, in particular the value of naturalness, meant transitioning from simple contrasts between captive and wild studies to more dynamic understandings of what constituted natural behavior and thus what made a field study a field study. Robert and Ada Yerkes's 1935 article in the *Handbook of Social Psychology*, for example, declared a simple and clear contrast between Solly Zuckerman's study of captive baboons and Carpenter's early field studies. The Yerkeses identified Zuckerman's 1932 study in the London Zoological Gardens' "Monkey Hill," made infamous by dozens of baboons living there and killing each other in a protracted fight, as exemplifying the failure of captive observations to capture any sense of natural primate behavior. Accessing such natural phenomena required work in the field, and the Yerkeses specifically praised Carpenter's Barro Colorado study of howler monkeys for demonstrating the potential for fieldwork to provide "exceptional inclusiveness and intensiveness."[5] For the Yerkeses, Carpenter's howler study allowed natural behavior to be observed and scientific rigor established by systematic and repeated observations. These characteristics may have made Carpenter's work lab like, but the Yerkeses understood his studies as starkly different from, and indeed far more valuable than, studies of captive primates like Zuckerman's now notorious "Monkey Hill" research that failed to shed light on natural primate behavior.

A closer examination of Carpenter's fieldwork reveals a far more complex understanding of naturalness than implied by the Yerkeses' black and white depiction of captive and wild studies. In reality, many of Carpenter's methods existed in a gray area, a zone where the meaning of naturalness was flexible and dynamic. In his 1930s field studies at Barro Colorado and during the APE, Carpenter's field methods certainly centered on minimizing the impact of the observer to ensure the behavior observed was natural. This characteristic of Carpenter's work was very much of the field. And yet, some elements of Carpenter's early primate studies reveal an ambiguity in his understanding of what constituted "natural behavior." For example, during the APE, a small number of primates were caught and kept captive. These captive specimens were sometimes used to demonstrate contrasts between abnormal and natural behavior and sometimes used to confirm observations made of wild, free primates.[6] Such confusion regarding what constituted naturalness only increased upon his return from the APE, when Carpenter set about establishing a primate colony on an island in Puerto Rico.

Carpenter's complicated understanding of naturalness pervaded twentieth-century studies of primates and indeed animal behavior studies more broadly. During this period, field researchers were increasingly applying experimental practices and thus modifying aspects of the behaviors observed. However, they remained tied to the fundamental assumption that the scientific value of fieldwork lay in its ability to do what captive studies could not—to see and analyze animals behaving naturally. Carpenter and others responded to this situation by modifying the very meaning of naturalness, creating new definitions and indeed a new vocabulary to simultaneously describe and validate the kinds of animal fieldwork being done in the 1920s to 1960s.

Naturalness at the Intersection of Biomedical and Behavioral Research: Cayo Santiago, Puerto Rico

Instead of basking in the glow of his methodological successes, after returning from the APE Carpenter began to focus on the limitations of fieldwork. He was particularly frustrated by what he perceived as the lack of efficiency that characterized field studies. He wanted to collect more data in less time, so soon after his return from the APE Carpenter joined a small group of researchers planning a semi-natural site for primate research on a Puerto Rican island called Cayo Santiago.[7] Aided by tattoos, feeding towers, and a confined area of land, the island colony would enable easier and quicker behavioral observations while also serving as a breeding colony for biomedical experiments. These two aims for the island—to observe natural behavior while generating subjects for biomedical experiments, some of which were performed on the island itself—could easily be seen as conflicting goals. In Carpenter's eyes, however, practical realities trumped any static view of naturalness, and he continued to define the primates' behavior on the island as natural despite the artificiality of many aspects of the colony.

Importing primates from India met demand for rhesus macaques for medical research in the United States during the first decades of the twentieth century. By the late 1930s, Carpenter and others were deeply concerned about whether India could continue to supply enough rhesus macaques for scientific research, not because of the risk of extinction but because of increasing difficulties associated with capturing and shipping the primates to the United States.[8] The quasi-sacred status of primates in India and the activities of both American and Indian branches of the Society for the Prevention of Cruelty to Animals both made it increasingly difficult to export primates from India.[9] Concerns about animal welfare intersected with a surge in American demand for primates for research, which increased from 12,992 in 1936 to 15,851 in 1938.[10]

Carpenter believed that meeting U.S. demand for experimental subjects required creating a reproducing colony of rhesus macaques. He embarked on the Markle-Columbia Primate Expedition in 1938 to locate and transport monkeys from India to establish a colony on Santiago Island. George W. Bachman, director of the School of Tropical Medicine in San Juan, and Earl T. Engle and Philip Smith of the College of Physicians and Surgeons, worked with Carpenter to secure funding from Columbia University, the Mary and John B. Markle Foundation, and the School of Tropical Medicine, Puerto Rico, for this expedition to Asia. Meanwhile, Bachman, in collaboration with M. I. Tomilin of the Philadelphia Zoological Park, prepared Santiago Island for the primates' arrival.

Working with two animal dealers identified simply as "Mr and Mrs A. W. Chater," Carpenter made an unsuccessful visit to the Jardin des Plants in Saigon and to three pet shops in China in search of gibbons. His search for macaques in India was far more fruitful and, after testing them for tuberculosis, Carpenter transported four hundred macaques to Santiago Island. Initially, a small group of gibbons, formerly captive at Columbia University, accompanied the macaques. After several attacks on humans, including one by a gibbon called Chipito who badly bit a woman, the gibbons were kept caged on the island before being sold to various zoological gardens in the United States.[11]

Buchman, Engle, Smith, and Carpenter identified several objectives for the Santiago colony, including attempting to establish a breeding colony for use in "multi-disciplinary scientific research and especially biomedical research" and studying primate behavior in a free-ranging environment. They also hoped to conduct "surgical and physiological interventions" and compare operated-on and normal individuals. Pursuing these goals required the development of methods in "primate husbandry," including how to use nutrition and disease control to produce healthy specimens that could be "harvested" for biomedical experiments.[12] The absence of the flora present in the macaques' native habitat, for example, meant the researchers had to install initially nine, and later twenty, food stations to provide the island's primate residents with Purina Chow, a dog food Carpenter identified as suitable for the primates. These feeding towers, combined with hand feeding, resulted in the macaques becoming "semi-domesticated." The macaques' inner thighs were also tattooed, allowing for individual recognition by a number, although names were also used for some of the primates. Study groups were identified by name. "Scarface Group" and "Shaggy Group," for example, were named after their respective alpha males.[13]

Carpenter believed that fieldwork on Santiago Island would be far more efficient than his previous field experiences and created an ambitious list of research goals that included inducing drug addiction, researching estrus and sexual behaviors including homosexuality, and studying the effects of stimulating isolated gibbons with recordings of vocal patterns.[14] Ultimately, the two articles that resulted from the two months Carpenter, accompanied by his wife, Mariana, spent on the island in 1940 focused on reproduction and sex, with the other plans failing to become a reality. The first article detailed what Carpenter identified as the "normal" sexual activities of a female macaque, including sexual swelling, presenting, and copulation. Overall, Carpenter depicted the female rhesus macaque as exhibiting significant sexual assertiveness during estrus. He ends with the summary: "the sexual hunger of the female and her capacities for copulation during her estrous period greatly exceed that of any single male during an equal time. A single estrous female may satiate, entirely or in part, several vigorous males; although when a period of such intense sexual activity terminates, the female may show extreme fatigue bordering on exhaustion."[15]

In his second article on his Santiago research, Carpenter turned his attention to homosexual, autoerotic, and what he termed "non-conformist" sexual behaviors. After detailing four case studies of female homosexual behavior, Carpenter summarized the behaviors of those in the "masculine" and "feminine" roles, pointing out that such "homosexual behavior may occur along with, preceding or following normal heterosexual behavior."[16] No such case studies were included in the section on male behavior, which stated that male-male mountings were an expression of "friendly" relations. The concluding paragraph of this section returned to "other kinds of deviate sexual behavior in female Rhesus monkeys" with an account of two females who forced themselves on disinterested males.[17] This paragraph was further supplemented by a whole section on "Non-Conformist Behavior of Estrous Females," which defined nonconformist as seeking out "foreign" males, or males outside their group, for copulation. With such seemingly sexually aggressive females, it is perhaps not surprising that a male rhesus macaque on Santiago, captured in a mesmerizing photo in *Life* magazine in 1939, was described as a "misogynist" who disliked females so strongly that he leapt into water to escape them (see figure 15).

Figure 15. This photo was included in *Life* magazine's January 16, 1939 issue with the caption "A Misogynist seeks solitude in the Caribbean off Puerto Rico." (Photograph by Hansel Mieth; Collection Center for Creative Photography; ©1998 Center for Creative Photography, the University of Arizona Foundation)

Carpenter's articles on his Santiago Island research identified "almost continuous, daily and systematic observation" as the method for drawing conclusions about macaques' behavior. In fact, Carpenter also applied a range of experimental practices while working on the island. For example, he conducted an experiment involving the removal of an alpha male to observe the impact on the group's dominance hierarchies. Such an experiment mimicked an occurrence that happens in many unmanipulated primate groups. The fact that the phenomena occurred in nature allowed Carpenter to understand the macaques' responses to this artificially created situation as natural, despite the artificiality of the stimulus provoking the behaviors. Such experimental interventions drastically increased what Carpenter referred to as the "efficiency" of fieldwork. Rather than waiting months or years for a change in social hierarchy, he was able to create the situation to suit his schedule.

Yerkes did not share Carpenter's apparent ease in accepting a natural interpretation of the behaviors observed on Santiago Island. In a generally positive report for the Social Science Research Council about the value of Carpenter's work on the island, he commented, "It is my opinion that 'the value and feasibility' of the undertaking turn primarily on the principle features of the island habitat. They are, I suspect, very unusual for *M. mulatta*, and social behavior may turn out to be correspondingly divergent from the normal or typical."[18] Yerkes was uncomfortable with the characteristics of the island, including the climate and flora, being viewed as representative of the primates' natural habitat. But despite Yerkes's apparent clear-cut understanding of naturalness, he also at times embraced a dynamic definition of what constituted natural behavior. Yerkes's shifting understanding of naturalness is best revealed when following him on what he called the "gorilla trail," a quest undertaken in 1950 to find and study what he believed to be the most difficult ape to observe: the gorilla.

Naturalness on the "Gorilla Trail"

The meaning of naturalness became exceptionally muddy where gorillas were concerned. The gorilla was understood to be shy and elusive to the point that field studies of its behavior were thought to be nearly impossible.[19] The methodological obstacles encountered by Bingham during his 1929 study of gorillas in the Belgium Congo had done much to reinforce this reputation.[20] The lack of field observations of gorillas in the wild compelled

Yerkes, accompanied by his wife, Ada, to dedicate much of the latter part of his career to what he termed, in a nod to M. H. Bradley's 1922 book of the same name, the "gorilla trail."[21] Setting out in 1950, Yerkes documented sites where captive gorillas could be observed for scientific study, a quest that took him to the San Diego Zoo, the New York Zoo, and circus winter quarters in Florida to census gorillas in captivity in the United States. He published an article in 1951 that included a table detailing where captive gorillas could be found and their names, sex, and age.[22] While on this "trail," Yerkes puzzled over topics as diverse as examining a baby gorilla's reaction to a red ball being bounced in front of her to whether a gorilla/chimpanzee hybrid would ever be possible.[23] Ultimately, however, Yerkes refrained from discussing such topics in his published papers.

Yerkes's 1951 article highlighted how little was currently known about the gorilla and pleaded for more studies of this intimidating yet shy creature. Captive gorillas were almost always the property of zoos or circuses rather than research institutions. Scientists who desired an end to the enduring dearth of knowledge concerning gorillas would need to follow in Yerkes's footsteps, exploring alternative sites for scientific study. Drawing on his own experience working with Congo in the 1920s, and his 1950 gorilla trail, Yerkes stressed that scientists could effectively work with owners of captive gorillas, often to the benefit of both parties. Through a combination of the "diversion, exercise and novel situations provided by the scientist" and "diary records of growth and development" it was possible for the investigator "to become an important asset instead of a nuisance in the zoo or circus set-up."[24]

During his time on the gorilla trail, Yerkes established a mutually fruitful relationship with Belle Jennings Benchley (1882–1973) and the San Diego Zoo (see figure 16). Trained as a teacher, Benchley was originally employed by the zoo in 1925 as a bookkeeper. She so impressed her coworkers that she quickly became executive secretary, essentially serving as the zoo's director until her retirement in 1953.[25] Benchley never held the official title of director although she was later referred to as "director emeritus."[26] Zealous about all her animals, Benchley had a special fondness for primates, and she devoted one of her three books to her experiences with them.[27] Although running a large zoo required a break from gender norms, her femininity, in the form of her maternal attachment to the animals, permeated the pages of her books, with one bearing the title, *My Animal Babies*.[28] Benchley was a critical component in the growth and success of San Diego zoo, including assisting Joan Morton Kelly, a psychologist from Pennsylvania State University and

Figure 16. Belle Jennings Benchley. (Courtesy of the San Diego History Center)

a former student of Carpenter's, in establishing a children's zoo within the grounds. In fact, Benchley discussed this enterprise with Ada Yerkes, and it was Robert Yerkes and Carpenter who suggested Kelly as a candidate for running the children's zoo.[29]

It had long been Benchley's dream to have a gorilla "family" at the zoo, and after the two adult gorillas that the zoo had acquired died, San Diego purchased three young gorillas for the children's zoo: two females named Bouba and Bata and one male named Albert.[30] Martin and Osa Johnson caught the gorillas while filming in the Albert National Park in the Belgium Congo. The nature reserve had strict rules preventing the collection of live specimens, but the Johnsons had special permission to capture one gorilla. Instead, they caught three and found themselves in "hot water" upon their return to the United States.[31] Benchley was aware of the less than ideal way the gorillas had found their way to American soil, but felt that, now they were in the United States, the gorillas had to go somewhere, and she was confident the zoo would be able to provide them with a comfortable home.

In 1950, Yerkes visited the San Diego Zoo to study the behavior of Bata, Bouba, and Albert. Yerkes's personal interest in the gorillas involved their ability to solve experimental problems, thus continuing his career-long interest in primate intelligence. Yerkes's work also extended to the practical demands of zookeeping. Collaborating with Benchley and Joan Kelly, Yerkes helped to train the gorillas to be more presentable to the American public. In tune with the pervasive domesticity of the 1950s, he worked on stopping the primates from eating their own feces and increasing their general obedience with the hope of enabling a "dinner act" for the zoo, an achievement realized by 1952 (see figure 17).[32] To enable the continuation of this training process once Yerkes had left, the scientist typed up a list of strategies for the zoo staff, specifically for Edalee Orcutt, the caretaker and observer of the gorillas. Although these guidelines focused on forging good manners for the zoo's domestic-inspired acts, the final step in the training program pointed toward preparing the primates for "systematic tests under controlled conditions." Yerkes suggested that these should be delayed for "several months," or until the gorillas were about three years old.[33]

Yerkes's study was not the first attempt at studying gorilla behavior at the San Diego Zoo. Harold Bingham also planned to research gorillas there. However, Carpenter's own research displaced this study. Carpenter was awarded $500 from the Social Science Research Council and spent the summer of 1934, following his fieldwork in Barro Colorado, studying gorillas at San Diego Zoo.[34] During this research, Carpenter relied on observation, making only one brief, failed use of experimentalist techniques. His article resulting from his time at the zoo notes that recordings of vocalizations would be useful, but he does not appear to have used such equipment himself.[35] Nor did Carpenter's publication address the issues of naturalness involved in such a study of captive animals, with only one line in the article noting that the behaviors observed might be "atypical."[36] His statement may well have been the result of a revision suggested by Yerkes in 1936. Writing to Carpenter, Yerkes cautioned him against generalizing statements made in reference to captive groups to gorillas more widely, commenting that "after all, two males, caged together, constitutes a very exceptional social group, and your results may be correspondingly exceptional."[37]

Nevertheless, at times, even Yerkes adopted an ambiguous understanding of what constituted natural behavior. Faced with a continuing dearth of field observations of gorilla behavior, Yerkes turned to more unorthodox sources of information on gorillas' social habits. In 1954, Yerkes was tremendously

Figure 17. Newspaper clipping featuring the gorilla dinner act from *Evening Tribune*, San Diego, California, March 15, 1952. (Courtesy of the San Diego History Center)

excited to learn of 2,000 feet of film footage of a group of gorillas in the wild. This treasure trove of gorilla behavior was produced during the filming of *Mogambo* (MGM, 1953), set in Africa and starring Clark Gable, Ava Gardner, and Grace Kelly. This was all the information Yerkes originally received about the film footage; he immediately developed a list of questions, including whether the gorillas were confined and, if so, how and whether any of the filmed behaviors were forced.[38] Thus, movie footage became a potential source of data for scientific research while the film itself became an artifact representing the cultural construction of gorillas and adventure in Africa.

Disappointed that he could not have been there for the original filming, Yerkes wrote to a Mr. Rubin, of MGM: "It would have been an absolutely unique opportunity to observe individual behavior for widely differing ages and also social relations and group behavior. Thus, within a week, I might have been able to learn more about the psycho-biological traits of this relatively little-known and extraordinarily interesting ape than is available in the world's literature."[39] In response to his questions, Rubin informed Yerkes that the primates were temporarily enclosed in a large fenced area and thus not completely free ranging.[40] Nevertheless, Yerkes remained excited about the possibility of seeing the 2,000 feet of gorilla footage. He traveled to Washington, D.C., to see the footage along with Coolidge and Leonard Carmichael, who was associated with the Smithsonian Institution. This was a long trip for a seventy-eight-year-old man, and Yerkes suffered a heart attack shortly after viewing the footage. His ill health continued and, although his viewing of the footage had motivated him to put plans in place by 1955 to produce educational and research films of gorillas, he was unable to realize them.[41]

Yerkes discussed the naturalness of the filmed gorilla behaviors with his friend and colleague, Benchley. Aware of the enclosed nature of the gorilla group, Benchley watched the film in California and commented to Yerkes that "while the situation was . . . not a perfectly natural one, the reaction to such a situation was certainly perfectly natural,"[42] a quote that expresses the complexity of any definition of what constituted natural animal behavior. Here, Benchley saw the gorilla's reaction as natural despite the artificial nature of the stimulus. The lack of field observations of gorillas and the belief this failure would continue led Yerkes and others to lower their standards for what constituted natural behavior and to settle for the most natural observations they believed were attainable.

Two of Yerkes' younger colleagues from the San Diego Zoo, Kelly and

Ken Scott, interpreted the film more critically, recognizing the expressions and behaviors as forced. Scott commented that the "worried look" of the gorillas "may be typical behavior . . . under harassment." Here, Scott was referring to the fact that humans had forced the gorillas into a fenced enclosure. Similarly, Kelly wrote in her report on the film that the "gorillas, though seemingly wild, are corralled and controlled within the area and from their behavior are very much aware that something is going on that they do not especially care for."[43] Nevertheless, both concluded that it was the best footage of gorillas in their native habitat they had seen. This episode demonstrates the difficulty in determining the naturalness of primate behavior while also suggesting an increasing awareness of issues of naturalness among a younger generation of primatologists. Such complex understandings of naturalness were not limited to Carpenter and Yerkes, or to primate research for that matter. A number of American scientists working with various animal species during the twentieth century were struggling with the concept of naturalness.

The Meaning of Naturalness: Looking beyond Primatology

By the 1940s, explorations of the concept of naturalness were taking place in a range of scientific conferences and symposia. For example, the New York Zoological Society's Committee for the Study of Animal Societies under Natural Conditions, which was formed in 1948 and composed of a diverse group of field biologists, provided a gateway through which to explore the issue of naturalness in animal behavior studies more broadly. Members included J. P. Scott, who was the committee's secretary, T. C. Schneirla, and J. T. Emlen and N. E. Collias, all ethologists working at the University of Wisconsin.[44] The committee discussed studies of animal behavior conducted under "natural conditions," although in reality the locations of the studies ranged from Barro Colorado, to Jackson Hole Wildlife Park and the Bronx Zoo, with seemingly little concern about the varying degrees of naturalness that characterized the sites.[45]

The 1948 meeting resulted in a special issue of the *Annals of the New York Academy of Sciences* two years later. Despite coming from diverse specializations within animal behavior studies, all of the committee members who contributed to the volume had serious concerns about the lack of status awarded to field research. Field biologists, such as Schneirla of the Department of Animal Behavior at the American Museum of Natural History, for example, argued that the observer in the field was not simply a "watcher" of nature but a true investigator.[46] Thus, the observer was a "participant" in the investigation not simply a "spectator."[47] As a result, the field biologist should apply, in Scheirla's words, an "experimental attitude" when conducting science in the field.[48] It is interesting to note that Schneirla supported his claim that fieldwork and researchers were scientifically rigorous by citing Carpenter's studies, specifically his work on gibbon behavior during the APE, as an excellent example of the successful application of repetition to field observation.[49]

Despite praise for Carpenter's fieldwork, his own contribution to the volume focused on the lack of status for field biologists. Four years later, in 1952, he continued to complain about the status of field science and its continuing reputation for being "qualitative rather than quantitative, general rather than specific, incomplete rather than complete and unsystematic rather than systematic."[50] Carpenter and others strongly believed the only way to improve this situation was to increasingly apply experimental practices in the field, despite the fundamental value of field research resting on the "naturalness" of the behavior the researchers observed and analyzed.

Embracing experimentalism while preserving the value of naturalness was not limited to primatology but in fact characterized a diverse range of animal behavior studies in the 1930s and 1940s. Ethologists John Emlen and F. W. Lorenz, for example, used sex-hormone pellets in their 1942 study of "wild-bird" behavior. The main advantage of using pellets, rather than injections, was the minimal disturbance to the birds created by the observer, a feature of the study stressed throughout their report.[51] The extent to which the observer influenced the behaviors observed had long been a basic criterion for naturalness, and thus this characteristic of the pellets allowed Emlen and Lorenz to remain closer to the naturalist ideal of observing undisturbed animal behavior. This study was one of many in which Emlen used experimentation, for he was a strong supporter of the integration of experiment into field studies. Writing as a member of the Committee for the Study of Animal Societies under Natural Conditions for the *Annals for the New York Academy of Sciences,* he commented that the traditional naturalist approach remained "basically unchanged" with new techniques "widening the scope of field studies." He added: "modern scientific standards have placed new demands for accuracy and caution. Experimental procedures have been introduced capable of measuring and analyzing phenomena which formally were reported only in general descriptive terms."[52] For Emlen, the incorporation of experiment into fieldwork was unproblematic.

A more familiar example of a researcher who combined experiment with the study of natural behavior is Niko Tinbergen, who began conducting simple field experiments during his doctoral research on *Philanthus* (the bee-wolf) in 1932 and started laboratory studies of sticklebacks by the mid-1930s. Tinbergen continued to study *Philanthus* and sticklebacks through the 1930s while also extending his interest to other organisms like the grayling butterfly. By 1946, Tinbergen was simultaneously exploring displacement digging in sticklebacks in the laboratory and the ability of aviary-raised jays to recognize insects in the field. Ultimately, Tinbergen established robust research programs in the laboratory and the field, often switching between scientific spaces depending on the seasons, while spearheading publication of *Behaviour*, a new journal for animal behavior studies, and writing ethology's first textbook, *The Study of Instinct*. For these achievements and others, he is identified as one of ethology's founding fathers.[53]

Warder Clyde Allee is another biologist who sought to combine experiment and field studies during the 1930s and 1940s. Focusing on group behavior and its physiological effects, Allee was part of the animal ecology program at the University of Chicago. He studied organisms ranging from isopods to birds and explored topics such as dominance hierarchies and cooperation in birds and the impact of hormone injections on domestic mice. It was Allee's "hope that this research would bridge the widening gap between the laboratory and the field, between the physiological methods of individual ecology and the descriptive natural history that characterized community studies."[54] It was only later, as the field of ethology matured, that Allee's work on pecking order received criticism for failing to recognize the impact of captivity on the dominance relationships of pigeons.[55]

The use of semi-controlled sites and experimental practices to cause varying degrees of manipulation to animal behavior became increasingly common from the 1930s to the 1950s through the work of Carpenter, Emlen, Tinbergen, and others. During this period, individuals such as Yerkes at times recognized clear distinctions between natural behaviors observed in the wild and artificial behaviors observed under captive conditions, while at other times, such as on the gorilla trail, understanding naturalness to be a more flexible concept. Investigators also continued to stress the importance of the observer not modifying the animals' behavior. The mounts, fences, and artificially implanted hormones could affect the animals' behaviors as long as their responses to such devices were not modified by the presence of the observer. In this way, investigators could use experimental practices and manipulate behaviors while still understanding their methods to be essentially true to the naturalist ideal of observing undisturbed behavior.

Natural Becomes a Troublesome Word

Applying experimentalist practices while continuing to use naturalness to validate fieldwork became increasingly problematic during the mid-1960s. A group of psychologists studying animal and human behavior, including that of primates, held two symposia, one at the University of Michigan and one at an American Psychological Association conference, in an attempt to reconcile experiment and the study of natural behavior. The purpose of the meetings was to discuss the meaning of the term *naturalistic* and to examine the relationship between naturalistic studies and experimentalism. Like the 1948 New York Zoological Society's Committee for the Study of Animal Societies under Natural Conditions, the contributors to the 1966 symposia believed that work in the field needed to be defended from the prevailing view that real science took place in the laboratory. For example, Edwin P. Willems, a psychologist at the University of Kansas, and Harold L. Rausch of the Department of Psychology at the University of Michigan, commented in the introduction to their edited volume based on the symposia: "Given our common disposition to defend a person's right to do the kind of research that he finds interesting and worthwhile, there is still a tendency, just as common, to view naturalistic research as a preliminary, early-stage, or even birdwatching, type of activity."[56] The context created by such concerns about status and recognition was intimately entangled with the view of experimentalism represented in the volume.

The contributors' perspectives ranged from arguing that naturalistic methods depended upon the absence of intervention to others who saw no clear distinction between observation and experiment. However, the vast majority of participants advocated for the integration of experiment with naturalistic research. William Mason, who worked at the Delta Regional Primate Research Center and had conducted a census of the macaques on Santiago Island with Carpenter in 1959, summed up this attitude in a review of the volume: "All seem to accept the necessity for manipulation and control but recognize that whenever the scientist arranges situations to suit his purposes, he thereby alters—or perhaps 'distorts' would be a better term—the very phenomena he seeks to measure."[57] Thus, the naturalness of the behaviors observed was compromised by the use of experiments in the field, however contributors hoped experimentalism would continue to colonize field studies of animal behavior.

The contributors embraced experimentalism by redefining the meaning of naturalistic research, identifying the practices used, rather than the behaviors observed, as lying at the heart of such an approach. As Mason commented, "naturalistic research is more a question of *what* the investigator does than *where* he does it."[58] A definition founded on practice provided a means with which to reconcile diverse views on the relationship between experiment and the study of natural animal behavior. Such a definition also meant that manipulation could be used and the behavior still be recognized as natural as long as the manipulation was limited to the minimum required to study the chosen question.

Willems' article for the volume advocated the use of a "two-dimensional space" in which laboratory and field studies were plotted in terms of the extent to which the environment and behaviors observed were manipulated. He argued that "the descriptive space avoids arbitrary dichotomies, and suggests that the manner in which the investigator functions in the process on generating data—and therefore what is often called the degree of naturalness—falls on a continua whose various points are so difficult to differentiate that they cannot be defended with finality and rigor."[59] As such, the conceptual model achieved two goals, creating a continuum to encourage methodological reconciliation between laboratory and field while redefining naturalness in terms of practice. According to Willems and the authors of many other papers from the symposia, any definition of this "troublesome word" in terms of behavior was both circular and divisive.[60]

Emil W. Menzel of the Delta Regional Primate Research Center provides a further example of the move away from the study of natural behavior, undisturbed, and toward the integration of experimental practices into naturalistic research. Highlighting the role of terminology in the creation of legitimization, Menzel reflects somewhat humorously on renaming the wilderness in which he studied: "In the hope that I can make field work scientifically respectable, I have considered patenting the tree as Menzel Jumping Stand, the river as Tulane Obstruction Apparatus, and the jungle as Delta Primate Center General Test Apparatus. These situations have already had several million years of standardization as reliable and valid tests of primate behavior, and who would challenge such a psychological tradition?"[61] Menzel went on to argue that naturalistic and experimental methods "are intimately related and necessary to each other."[62] In this sentiment, Menzel is representative of the vast majority of contributors to the volume. Unlike the other authors, he openly locates his argument within a progressive narrative, identifying naturalistic research as sharing the destiny of alchemy and astrology: doomed for extinction, only to be replaced by the next stage of science's evolution, in this case, experimental studies.[63]

As the field became increasingly characterized by experiment, practice rather than place became the crucial component of a definition of naturalness. Thus, practice became the essence of the meaning of natural. As contributors to the symposia on naturalistic research argued, the term *natural*, when defined according to the behaviors observed, had simply become too ambiguous and, in turn, too controversial. The proceedings capture both the divisions that characterized the field and lab from the 1930s into the 1960s and the culmination of these tensions in a cry for reconciliation in the form of a redefinition of "natural" in terms of practice. By using practice as the basis for determining the naturalness of observed behaviors, researchers had a flexible, nondivisive, term that allowed them to use new experimentalist techniques while continuing to identify their research as the study of natural behaviors. Thus, "natural" could hopefully cease to be the "troublesome word" that had resulted in such controversy within the biological and social sciences.

During the first half of the twentieth century, access to certain species and continuing attempts to demonstrate primate studies were real science, distinct from amateur's tales, resulted in definitions of naturalness being bent to adhere to scientific standards of control, repetition, and efficiency. In this changing climate, Carpenter and others turned to experimental field practices. Some of these experimental practices reflected laboratory experience while others continued and modified traditional naturalist techniques. Regardless of their epistemological origins, these techniques were developed as part of researchers' pursuit of manipulative practices that allowed observed animal behavior to still be deemed "natural."

Gradually primatologists conceded that natural behavior did not require animals to be unaffected by the presence of the observer. Instead, behavior retained its naturalness even when the observer was actively manipulating the behavior being observed. However, a clear distinction was formed between an observer's manipulation of the animal's response and the behavior's stimulus, the former being unacceptable as it broke the naturalist's rule of observing undisturbed behavior, the latter being an inevitable consequence of the application of experimental techniques. The behavior of pigeons in response to implanted hormone pellets was understood as "natural" because Emlen and Lorenz had modified the stimulus of the behavior, not the response itself. Likewise, Yerkes and Benchley interpreted the behaviors of the entrapped gorillas in the film *Mogambo* as natural because the fear responses were natural, even though such confinement was not.

A dynamic understanding of naturalness enabled field biologists like Carpenter to use experimentation in the field while maintaining an identity as field investigators whose work was scientifically valuable because, unlike their laboratory counterparts, they studied natural animal behavior.[64] Furthermore, as the 1966 symposia demonstrate, redefining naturalness according to practice created a sliding scale that held the potential for uniting lab and field-based studies of behavior.

In the 1920s and 1930s, when the field needed to be promoted as scientifically valuable, Yerkes and Carpenter opted to emphasize the differences between natural behaviors observed in the field and the artificial behaviors seen in captive animals, whether in the lab or zoological park. As pressures from within and outside primate studies increased, such as concerns over scientific status and growing demand for primates for biomedical research, the meaning of "natural" became more flexible. Doubtless this adaptation was spurred by practical considerations, including Carpenter's career goals and the need for primate studies to respond to a changing social, cultural, and intellectual climate. It would seem that the development of primatology, and animal behavior studies more widely, into a recognized science, with the status and funding such recognition brings, demanded that rhetoric of "naturalness" be both contained and continuously redefined.

For a small group of young primatologists emerging in the 1950s and 1960s, the quest for observing truly natural behavior, the type only seen when animals are studied for prolonged periods in their native habitat, was far from over. For two of these researchers, George Schaller and John Emlen, the "gorilla trail" did not lead to brief visits to the cages of zoological parks but instead to an eighteen-month field study in the Congo. For others, understanding natural primate social behavior would come to necessitate multidecade, rather than multimonth, stints in the field and a recentering of noninterventionist, rather than experimental, practice.

5

Modern Primatology

The Emergence of Long-Term International Field Projects

Prior to the 1960s, researchers such as Robert Mearns Yerkes and Clarence Ray Carpenter laid the foundation for the science of primatology by adopting methods and techniques to allow field studies of primates to be accurate, objective, and controlled, while at the same time they emphasized the naturalness of behavior observed in the wild. At the heart of these early field studies were questions about primate biology and behavior, such as primates' estrous cycle and locomotion, how they cared for their young, and how they maintained social cohesion. Armed with an expanding arsenal of methodologies and technologies, early primatologists embarked on short stints in the field that were increasingly efficient at gathering information about primates.

The development of primate field studies stalled during the Second World War as scientists' attention and patrons' funds shifted toward practical wartime matters. When primate studies resumed in the 1950s and 1960s, primatologists' research was no longer based on data they collected during fleeting field expeditions but rather on sustained observations of groups of primates over multiple years. Indeed some of the projects that started in the 1960s generated decades of continuous observations. It is surprising that the gorilla, previously believed to be nearly impossible to observe in the wild, emerged as one of the first subjects to be successfully observed for an extended period.

An important milestone in the transition from short field trips to long expeditions took place in the late 1950s, when researchers from the University of Wisconsin undertook an eighteen-month study of gorillas in the Congo. Like many of the primatologists who had preceded them, John Emlen and George Schaller focused on the identification of individuals as well as their social behavior and vocalizations. However, they differentiated their work from earlier primate field studies by emphasizing increasingly rigorous standards for fieldwork, which necessitated longer stints in the field, observing a species previously thought to be impossible to see in the wild, and basing their conclusions on hundreds of hours of observations.

The generation of primatologists who dominated the field in the postwar years reinvented primatology by establishing the field itself as the central location for research, and in so doing fundamentally altered the very nature of modern primatology. Starting in the 1960s, Jane Goodall and Dian Fossey used long-term fieldwork to demonstrate the great apes were even more human-like than previously thought, and they became cultural icons by virtue of the nature of their research, the exotic locations of their field sites, and the visual appeal of both the scientists themselves and their primate subjects. These factors combined with new forms of media as well as conservation and animal activism to create a primate folklore for the modern age that I refer to as *pop primatology*. Sharing a number of characteristics with nineteenth- and early twentieth-century primate stories, popular accounts of primate studies in the postwar period also differed in important ways, including the unprecedented level of celebrity gained by both scientists and their primate subjects.

Amid the growth in pop primatology, a select group of researchers charted a new direction for primatology, one that continued a long-standing emphasis on observation and description but significantly increased the role of quantification and mathematics to study primate behavior. These researchers may have failed to attract the media's spotlight, but they lead the way in integrating emerging fitness and life history theories with long-term primate field studies.

Making the Impossible Possible:
Postwar Studies of the Gorilla in the Wild

In the fall of 1931, animal advocate Richard Sparks wrote to Harold Bingham, a former postdoctoral fellow of Yerkes, lamenting the popular press' long-standing mischaracterization of gorillas: "How long, oh, how long before a reputable publication will carry the truth of the gorilla."[1] Twenty years later, the world was still waiting for a successful field study of gorillas, and readers had to settle for the captive specimens found on Yerkes's gorilla trail. It was not until the late 1950s that scientists finally observed natural gorilla behavior during what became the first of many long-term field studies.

In the summer of 1958, the New York Zoological Society organized a meeting to discuss a potential African Primate Field Expedition. Attendees included Carpenter, Stuart Altmann (then employed by the National Institute of Neurological Diseases and Blindness), John Emlen (from the Department of Zoology at the University of Wisconsin), George Schaller (one of Emlen's graduate students), and Harold Coolidge of the National Academy of Sciences. After the meeting, Emlen submitted a proposal to the New York Zoological Society for a study of "The Ecology and Behavior of the Mountain Gorilla in its Native Habitat." When reviewing the proposal, Carpenter ranked it as "acceptable." His main concern was that the field expedition was to be composed of too few people, with too little training, and insufficient funds. He also noted his personal "distrust of the man-wife combination," despite the fact that his first wife accompanied him on his early field studies.[2] Ultimately, the New York Zoological Society, with support from the National Science Foundation, funded John Emlen, George Schaller, and their wives on what would become a yearlong study of gorillas in what is now the Democratic Republic of the Congo, then simply known as the Congo. After an initial six months spent exploring the most favorable site from which to conduct the field study, the Emlens returned to the University of Wisconsin. Schaller and his wife, Kay, stayed behind for another year to study gorillas in the Albert National Park within a region of the Congo called the Virunga Volcanoes. Harold Bingham had attempted, and failed, to systematically observe gorilla behavior in this same area as part of Yerkes's series of field studies in the 1920s.

In sharp contrast to Bingham's observations of distant, shadowy gorillas, Schaller successfully habituated several of the apes, allowing him to collect 466 hours of observations and publish extensive data on the population, ecology, and behavior of the mountain gorilla. Whereas Bingham led a group of porters and gun bearers into the jungle, Schaller observed the gorillas either alone or accompanied only by his wife. Using tracking techniques he learned from African guides, he made daily visits to "one or two of the ten gorilla groups that frequented the forest . . . for weeks on end, alone and unarmed, until the animals became habituated to my presence and I was able to observe their daily activity and behavior for hours at distances ranging from 5 to 150 feet."[3]

Schaller was particularly interested in observing fundamental and previously unknown aspects of gorilla biology and social life. Using information gleaned from studies of captive apes, he classified the wild gorillas according to age and, when possible, sex. Schaller described gorillas' bodies, in particular their faces, and included sketches of their nose markings that allowed for identification of individual gorillas. He also described gorillas' use of sight, smell, and hearing, as well as their body postures and styles of locomotion. After summarizing bodily functions like urination, hiccupping, and burping, Schaller spent several pages detailing gorillas' facial expressions and vocalizations. Sounds were described in words and represented visually by spectrograms. Turning to social behavior, Schaller looked at dominance, grooming, and sexual behavior and, similar to Carpenter, divided social interactions into male-male, male-female, female-juvenile, and so forth.

Schaller published his descriptions of gorilla behavior in his 1963 monograph *The Mountain Gorilla: Ecology and Behavior,* and he wrote a shorter book, *The Year of the Gorilla,* aimed at the general public a year later. Descriptions of gorilla behavior and habitat, supplemented by photos of the forest, gorillas, their nests, and Africans, dominated both texts. His scientific monograph also included several line drawings, depicting nose prints, knuckle and footprints, body postures, and home ranges. Schaller even included line drawings of cross-sections of gorilla dung to reveal information about gorillas' diet and size. The few graphs included in *The Mountain Gorilla* detailed the daily activity of gorillas, including when they slept, and spectrograms of gorilla vocalizations. Graphs were also used to represent the body weight of individuals according to age and sex. Such diagrams and graphs were absent from *The Year of the Gorilla,* which instead wove together travel narrative, diary entries, and scientific observation to transport the reader to the Virungas.

As the first prolonged field study of an African ape, Schaller's research garnered attention in both academic and popular quarters. One cartoon in the *St. Paul Pioneer Press* focused on the length of time Emlen spent in the field, suggesting that he might "go native" and be fired from his academic appointment at the University of Wisconsin. "The Amiable Gorilla," in a 1960 issue of *Sports Illustrated,* incorporated sections written by the journalist John O'Reilly with Schaller's own words and focused on the sharp contrast between French explorer Paul Du Chaillu's "exaggerated accounts" of gorilla behavior and Schaller's "mass of data on the life history and behavior of the wild gorillas" to demonstrate the tremendous shift in professional opinion about the nature of gorillas. In contrast to earlier accounts, Schaller depicted the gorilla as "an amiable animal" living "in peace with others of its kind and with the world around it." This ape was living a "life of leisure" and was so similar to humans that "all their actions seem like a parody of their larger-brained relatives." With over six pages of description, the article relayed the experiences of Schaller, "the young biologist," and "his pretty blonde wife Kay," with an emphasis on the length of the field study, the naturalness of the environment, and the "friendly coexistence" in which the Schallers, "alone and unarmed" lived with the wild apes.[4]

In reality, the Schallers were not alone in the wilderness. Two Africans, one employed as a "camp boy" and the other as a guard, accompanied them but go unnamed even in Schaller's *The Mountain Gorilla*. The book's acknowledgments did, however, capture the increasing complexity of primate fieldwork, with thanks extended to a host of funding agencies and colleagues in the United States as well as park wardens in the Congo. They also included an expression of gratitude "to the numerous Bantu and Batwa guides and porters, whom we employed at various times in Africa and whose help was essential."[5] The methods section of the book included discussion of the contributions of local people, highlighting the roles played by "natives" in locating gorillas and identifying the plants they ate.

Schaller described his African guides as excellent trackers. But, rather than rely on a permanent team of guides, he asked his Ugandan guides to teach him how to track. "Knowledge of tracking the animals by signs as a chewed piece of bark, a bent blade of grass, or a knuckle imprint is essential for any investigator," he explained, "for good forest guides are rare and none are dependable."[6] Despite his willingness to learn from the locals, Schaller's discussion of behavioral observations also includes negative depictions of Africans. Schaller cautioned that he "encountered not a single African who gave consistently reliable information regarding the behavior of gorillas."[7] Even behavioral observations provided by Schaller's guide, Reuben Rwanzagire, were "treated with reserve," despite Rwanzagire's knowing each gorilla individually.[8] In response to the apparent lack of validity of most indigenous knowledge, Schaller relied on his own observations and occasionally those provided by Emlen and a small collection of unpublished observations made by a few colleagues, including his wife, Kay.

With regard to his data, Schaller's decision to trek into the jungle either alone or accompanied only by Kay was clearly a productive one. Had he brought a small army of porters and gun bearers as his predecessors had, he surely would not have successfully completed the first systematic study of the elusive wild gorillas. However, his comments concerning the roles that local people played in his eighteen-month field study highlight the complex relationships that had always existed between Western primate researchers and the "natives" they encountered while in the field, be it on short expeditions of a few weeks or the increasingly long expeditions that were by now lasting several months.

Following in Schaller's footsteps, Dian Fossey studied gorillas in the Virungas from 1967 until her murder in 1985.[9] Her PhD dissertation, which was based on her work in Parc des Virungas, Zaire, and Parc des Volcans, Rwanda, provided answers to many of the questions Schaller had raised about gorilla immigration and emigration, particularly about how females transferred between groups. Through close study of a number of habituated groups, Fossey identified "home groups" as those composed of individuals born into cohesive and relatively stable groups and "transfer groups" that were formed when a female joined a silverback male to form an unstable union. Fossey's prolonged observations of group interaction and migration contributed to understandings of primate reproductive behavior, including the phenomena of infanticide, knowledge about which emerged during the 1980s.[10]

The techniques that Fossey developed allowed her to habituate gorillas and be privy to their movements, vocalizations, and social behaviors. She mimicked "contentment vocalizations," for example, as well as eating, grooming, and chest beating. Although others had occasionally used imitation of vocalizations, Fossey applied the technique of mimicry more extensively than had her predecessors and described the method in her popular *National Geographic* articles. In her 1970 article, "Making Friends with Mountain Gorillas," for example, Fossey outlined her imitation techniques before commenting that "admittedly these methods are not always dignified. One feels a fool thumping one's chest rhythmically, or sitting about pretending to munch on a stalk of wild celery as though it were the most delectable morsel in the world." As a result of such popular articles, Fossey became well known for making prolonged observations of gorillas by acting like one of them.[11]

Fossey's long-term study of gorilla behavior also involved extensive use of indigenous peoples and indigenous knowledge. Like Schaller and the generations of primatologists and explorers before him, Fossey hired local people as trackers and guides. As her research progressed, however, so too did her concern for gorilla conservation, and she ultimately also hired local people to patrol the park to deter poachers. Despite their contributions to both Fossey's research and her conservation efforts, the role of indigenous peoples as the hunters of gorillas, rather than their protectors, formed the basis for the popular articles and books written by and about Fossey. Contributions by indigenous peoples in Fossey's long-term field research is barely identifiable in her dissertation, and, while some individuals were mentioned for the help they provided Fossey in her widely read 1983 book, *Gorillas in the Mist*, local people in general are typically depicted as poachers.[12] Indeed, Fossey stated bluntly: "The Batwa are poachers pure and simple."[13]

In *Gorillas in the Mist*, Fossey provided detailed descriptions of poacher's traps and the death and injury of gorillas unfortunate enough to be caught in them. The death of Digit, a large male gorilla with whom Fossey had formed a special bond, was the most emotional and haunting of Fossey's accounts of poaching. Digit's "mutilated corpse" was found "in the corner of a blood-soaked area of flattened vegetation. Digit's head and hands had been hacked off; his body bore multiple spear wounds."[14] A graphic photo of Digit was included in *Gorillas in the Mist*; once seen, the image cannot easily be forgotten. After Digit was photographed and his corpse examined, Fossey buried him in her field camp. In time, the camp came to include an entire gorilla graveyard. Fossey saw this act as "a memorial in the hopes that the day won't come when there are only graveyards and memories in the mist."[15] After her murder in 1985, Fossey's own body was buried among her gorilla companions.

In addition to being killers of apes, poachers practiced *sumu*, or black magic, and frequently attempted to communicate death threats, administer poison, or gain control over Fossey's body and spirit.[16] Fossey responded to these tactics by resorting to imprisonment to deter poachers. After locating one of the men responsible for Digit's death, Fossey chased him with a gun and later, with the aid of her Rwandan field assistants—whom she referred to as "my Africans"—the man was "questioned" and "interviewed." Fossey asked two assistants to spend the night guarding the man because she "could not trust [her]self alone with that thing."[17] The long acrimonious relationship between Fossey and the Rwandan poachers led them to be suspected of her murder, which remains unsolved.[18]

Characterized as "active conservation" or what feminist science studies scholar Donna Haraway has termed "anarchist direct action," Fossey literally took up arms in an attempt to protect the primate subjects of her science.[19] She is not the only researcher to combine guns, activism, and conservation while conducting long-term animal studies. Iain Douglas Hamilton, who began studying elephants in Lake Manyara National Park, Tanzania, in 1965, used guns to protect his scientific subjects. Like Fossey, Douglas Hamilton conducted a long-term study focused on animals with complex emotional and social lives and, when attempts to lobby for elephant protections failed in 1981, he took up "G3 and AK-47 automatic rifles ... and organized a paramilitary operation to protect 160 elephants."[20]

Fossey's combination of gorilla science, conservation, and activism catapulted her to fame, attracting sufficient attention during and after her lifetime to warrant coverage in *National Geographic, the New York Times, Vanity Fair,* and *Vogue* in addition to popular books, a series of episodes of *Wild Kingdom,* and a motion picture staring Sigourney Weaver.[21] In popular representations, Fossey is consistently the hero, with the gorillas requiring her protection from Rwandans, who are depicted as ignorant poachers. The narratives and images render the Rwandan field assistants' contributions to Fossey's gorilla project completely invisible.[22] Unfortunately, this invisibility has continued to characterize popular depictions of primatology despite projects like the Gorilla Organization, which was formerly known as the Dian Fossey Gorilla Fund Europe, now being led by African conservationists.

Pop Primatology:
A New Primate Folklore for a New Age

Popular accounts of primate studies created a challenge for primatologists seeking to further separate themselves from the amateurs before them. Best sellers like *Gorillas in the Mist* and iconic *National Geographic* front covers created a genre of pop primatology in the 1960s and 1970s that focused on studies driven by observation and description and that were performed by women without scientific training, although both Goodall and Fossey eventually earned PhDs. Popular coverage blurred lines between formal scientific research, popular natural history, conservation, and activism, and it in turn created a complicated terrain for the primatologists who began their careers during these decades to navigate.

Pop primatology in the late twentieth century formed part of a larger trend, as a new sensibility about humans, their development over time, and their contemporary behaviors began to emerge in the biological and social sciences. In particular, after World War II and during the Cold War, people increasingly turned to studies of animal behavior in hopes of understanding human aggression and warfare. Ethologist Konrad Lorenz's *On Aggression*, which was published in 1966 and used animal behavior to examine human's apparent instinct for aggression, is the best-known example of this genre.[23] Such popular stories and texts, driven by anthropomorphism and dramatic description rather than quantified systematic data, attracted readers but they also raised epistemological problems for researchers of animal behavior.

Images of individual Western researchers in the jungle protecting apes from Africans formed the heart of popular representations of primate studies in the 1960s and 1970s. Primates' value in terms of conservation, their evolutionary connection to humankind, and the emotional connection between the animal subjects and human observers took center stage in both narratives and images.[24] Enduring public interest in primates may have helped maintain and increase funding streams for primatology, facilitating the creation of new sources of support for primatologists in the form of charity contributions and ecotourism, but it also perpetuated a fundamental misrepresentation of the science. By highlighting the length, naturalness, and wildness of primate field studies while focusing on a white Westerner supposedly alone in the jungle, most popular media accounts of primatology in the late twentieth century retained many of the central characteristics of the previous century's primate folklore. A key difference was that scientists, rather than hunter-adventurers, were now the authorities on primate behavior.

Alongside the Western scientists featured in the popular press were images and descriptions of charismatic great apes. Unlike the primate folklore of the nineteenth and early twentieth centuries, which oscillated between emphasizing primates' bestiality and their humanity, popular representations of primates in the second half of the twentieth century focused on their humanlike characteristics. Leveraging the results that were beginning to emerge from long-term field research, and in some cases laboratory studies, primatologists initiated a series of challenges to specific claims that had long been used to declare humans as distinct from other primates, such as tool use, warfare, and language. Throughout the second half of the twentieth century, primatologists demonstrated that we share each of these traits with our beastly kin, and the popular press quickly picked up on these scientific discoveries.

By this time, apes had been observed performing acts previously believed to be unique to humans, including using tools to fish in termite mounds and engaging in terrifyingly human-like acts of violence. Jane Goodall observed tool use by chimpanzees in Gombe, Tanzania. At first Goodall used this observation, combined with descriptions of their social structure, to declare chimps to be a lot like humans, only "nicer." By 1974, however, Goodall's romantic view of chimps' behavior was "shattered by division and war" among primate groups.[25] During a time when the public was particularly concerned about what propelled humans to be aggressive, Goodall observed acts of violence, some of which were "certainly the equivalent of human murder."[26] From how they nursed their young, kissed their companions, and killed their foes, chimps suddenly seemed to be a lot like us. Clearly, our best and worst qualities appeared in the animals we recognized as our closest evolutionary kin. Together, Goodall's observation of tool use and murder catapulted both her and her primate subjects to fame. She became a cultural icon representing conservation and women in science, and the public knew many of the chimps she studied by name.

Washoe, who was observed in captivity and was fluent in American Sign Language (ASL), also became a household name in the 1960s. Washoe represented apes' apparent potential to bring down yet another boundary between humans and primates: the ability to use language. Psychologists Allen and Beatrice Gardner studied Washoe at the University of Nevada during the 1960s. Roger Fouts, one of the Gardners' students, and his wife, Deborah, continued the project at Central Washington University. Both the Gardners and Fouts immersed Washoe in ASL.[27] Another scientist, Herb Terrace, expanded the study to another chimp named Nim at Columbia University. Shortly thereafter, Washoe and the scientists who studied her became embroiled in a controversy surrounding their methods for teaching and observing chimps signing with Terrace, declaring his own work and that of the Washoe project as significantly flawed. Terrace claimed that the chimps were imitating rather than creating novel, meaningful, structured sentences as Fouts, and initially Terrace, had first thought. Despite the controversy, or perhaps because of it, researchers such as Roger Fouts and primate subjects like Washoe remain well known, if not the cultural icons that Fossey and Goodall came to be. Fouts was quite famous by the 1980s, indeed his "international reputation" led him to be hired as a consultant for the 1984 film *Greystoke: The Legend of Tarzan, Lord of the Apes*, which featured athletes, acrobats, and dancers trained by Fouts in primate behavior and dressed in chimp costumes.[28] Ultimately, Washoe's celebrity status outshone her human observers; indeed her death in 2007 warranted an article in the *New York Times*.[29]

Pop primatology was particularly focused on the great apes and how they could be used as mirrors with which to see human nature. Fossey and Goodall, for example, each studied great apes and were specifically interested in their humanlike qualities. Similarly, captive studies focused on building bridges between humans and other apes by developing language in chimps like Washoe and gorillas like Koko. Psychologist Francine Patterson and Koko were featured on the front cover of *National Geographic* in 1978, for example, with a lengthy and heavily illustrated article about Koko's ability to have meaningful conversations.[30] Primates' similarity to humans held both the public imagination and motivated many primate researchers' work.

Beyond interest in the nonhuman primates, the fact that primate studies typically concentrated on white women seemingly alone in the wilderness increased popular interest in field primatology. *National Geographic* covers featuring Goodall were part of the collective consciousness of those who grew up in the 1960s and 1970s.[31] Primatologists like Mike Wilson, former Director of Research of the Gombe Stream Research Center, identified these stories as the catalysts that propelled many budding primatologists' career paths by creating a desire for adventure and a passion for conservation.[32] Images of female primatologists in popular magazines also created a reputation for primatology as a science open to women, and it continues to attract and retain many female scientists. In 1984, female primatologists were already sufficient in number to warrant an edited volume dedicated to their research, *Female Primates: Studies by Women Primatologists*.[33] By 2000, the number of women in primatology hit 62 percent, compelling primatologist Carol Jahme to publish a book examining the extent of women's involvement in primate studies.[34]

By the 1970s scientists like Goodall and Fossey were leveraging their popularity and their research on the humanness of chimps and gorillas to call for conservation of great apes. At the same time, concerns about captive apes began to emerge in the popular arena with Frederick Wiseman's documentary *Primate* (1974) making public many previously private practices of primate laboratories, including the use of electrodes in the brain to stimulate masturbation.[35] Several prominent newspapers initially reviewed *Primate* but the documentary faded from popular memory, probably due to the graphic nature of its images, which makes viewing it rather unpleasant.

In contrast, Fossey and Goodall became cultural icons in part due to the popular appeal of glossy photos of cute apes and white women in the jungle.

Beyond Pop Primatology: Scientists' Vision for the Future of Primate Studies

In contrast to the relatively small group of primatologists who became household names, several primatologists during the same time purposefully shied away from the limelight. Instead, they preferred to pursue methodologies and data that lacked the necessary sexiness to warrant discussion on the pages of popular magazines or TV shows. In particular, a select group of primatologists came together in the early 1960s to take stock of where primate studies had been and where it ought to be heading. Two edited volumes were published in 1965 that summarized the current state of the field, one by anthropologist and evolutionary biologist Irven DeVore entitled *Primate Behavior: Field Studies of Monkeys and Apes* and another called *Behavior of Non-Human Primates: Modern Research Trends,* which was edited by three scientists researching primates in the laboratory, Harry Harlow (renowned for his research on attachment between wire and terry-cloth mothers and primate infants) from the University of Wisconsin and Allan M. Schrier and Fred Stollnitz from Brown University.[36] Both volumes sought to summarize primatology's key achievements to date, new directions for the future, and spearhead greater discussion between laboratory and field researchers.

Devore's volume was the product of a nine-month "Primate Project" at the Center for Advanced Study in the Behavioral Sciences in Stanford. With funding from the National Institutes of Health, the "Primate Project" sought "to evaluate the present position and future possibilities of primate field studies," while determining future relationships between field and lab studies with the goal of exploring issues in psychology and psychiatry.[37] Schaller participated in the project for the entire nine months, while others such as Carpenter, to whom the edited volume was dedicated, made brief visits to the center at Stanford. Working conferences at the start and end of the project facilitated planning for the volume *Primate Behavior* with the goal of using it in the primate ecology classes that were beginning to be taught in the 1960s. The book thus represents a moment when researchers came together "to arrive at mutual understandings concerning the description and interpretation of primate behavior" and to create intellectual and methodological cohesion for future primate studies.[38]

In taking stock of primate studies in 1965, the editors reminded the readers in several sections of *Primate Behavior* that adaptive behaviors could only be seen in the wild, the number of field studies conducted to date was still small, and more research was needed to fill the significant gaps in knowledge that remained for many primate species. The importance of not interfering with primates and limiting the use of experimental practices in the field was emphasized, with methods such as *provisioning,* which was common in Japanese primatology, highlighted as a practice that "alters the ecological and behavorial patterns of the animals."[39] Thick description was a characteristic shared by all of the essays, although wherever possible graphs, matrixes, tables, and diagrams supplemented the descriptions with quantified data. This was particularly true for discussions of dominance, copulation, and spacing of individuals and groups.

Primate Behavior concluded with a discussion entitled "The Implications of Primate Research," which made a clear distinction between the "anecdotal era," when "the occasional observations of native, hunter, or traveler" could adequately describe primate behavior, and modern field studies. "Perhaps the most important conclusion of the field studies so far is that it is in fact possible to obtain quantities of reliable data. In marked contrast to the anecdotal era, the studies are now critical and cumulative.... Individual animals can be recognized and the role of personality in the social structure described. The complex interrelations of different species can be observed."[40] Here, DeVore took pains to demonstrate that studies of primate behavior were no longer limited to the unreliable information that characterized the "anecdotal era" and to declare all the things that modern primate research could accomplish. By framing his discussion in relation to the stories of hunters and travelers, DeVore demonstrated that modern primatologists still felt the burden created by the amateurs' stories that had come before them.

Although a goal of the "Primate Project" was to explore ways to bridge field and laboratory studies, little mention is made of the lab. It was left to Harlow, Schrier, and Stollnitz to provide a thorough review of the state of primate lab studies in 1965, a time that marked a transition from two main primate laboratories in the United States—Yerkes Laboratories of Primate Biology and the University of Wisconsin Primate Laboratory—to several in the form of government-funded Regional Primate Research Centers. *Behavior of Nonhuman Primates* eventually stretched to four volumes, spanning studies of learning, hearing, vision, affection, communication, operant conditioning, and the effect of exposure to radiation. One chapter was dedicated to field studies, with another examining the development of social behavior, which was written by British zoologist Robert A. Hinde, who supervised Goodall's and Fossey's doctoral dissertations.

It is ironic that two volumes that sought to bridge field and lab primate studies so concretely embodied the divide between the two approaches. Altmann reviewed both texts for *Science* and highlighted the schism they represented. He wrote:

> Many of those who were trained for laboratory research in psychology are adroit experimenters; some of them are masters at experimental design, statistical tests, mathematical models, checking on observer reliability, and other aspects of scientific methodology that are the weak points of most field work. The fieldworker, in contrast, sees primate behavior in the context of adaptions to cope with the problems that these animals face in their natural environment, a perspective which most laboratory workers are unfamiliar with. These surveys of the literature present us with some of the best of both worlds. Let us hope that subsequent editions will reveal that each group has learned from the other—indeed to such an extent that only a single unified review will be necessary.[41]

Looking forward, Altmann and others clearly hoped for a primatology that took the best from the field and the lab, combining sophisticated quantification with prolonged observation in the wild, and a focus on noninterventionist practices while recognizing the value of experimental techniques when the need arose. Most important, those researchers who sought to spend their careers in the field wanted to declare that the so-called anecdotal era was far behind them, if still weighed on popular and professional notions of their work.

Primatologists like Stuart and Jeanne Altmann were establishing their own field sites by the 1970s and seeking to extend quantification in field studies to include measures of observer bias, social behavior, and fitness. Reflecting developments in evolutionary theory, population genetics, and ecology, long-term fieldwork began to track organisms through the durations of their lives, creating opportunities for extensive data collection about the adaptive value of a range of behaviors and life events. The subjects they chose were savannah baboons, which are terrestrial and thus relatively easy to observe but lack the charisma or cuteness to become front-page news. The Altmanns studied them to determine the adaptive value of behavior, not purely as a means to understand our own evolution, and for the most part their work attracted little popular coverage.

The new questions and theories that propelled primatology in the 1970s demanded that field studies get even longer, extending from a few years to as long as decades. As early as the 1920s, local people had provided Western primate researchers with essential logistical support and knowledge that would otherwise have been inaccessible to them. However, when it came to what counted as valid behavioral observations, researchers drew a strict line between what Western and indigenous eyes saw. This long-established characteristic of the relationship between primatologists and locals came into the foreground as field studies began to extend beyond the purview of what leave a Western academic could take from his or her university or the length of funding they could receive from their patrons. This development also intersected with the emergence of a new generation of primatologists, some of whom became increasingly conscious of the colonial dimensions that inevitably impacted scientific endeavors that are led by Western researchers on African soil. For at least one long-term primate study, the Amboseli Baboon Project, the combination of these factors led to a radical rethinking of the role indigenous peoples should play in modern primate research.

6

Inclusion and Indigenous Researchers

The Africanization of the Amboseli Baboon Project

Several current primate field studies have endured for twenty-five years or more. Unlike many researchers in the 1940s and 1950s, the scientists working on these projects adhere to non-interventionist practices whenever possible, believing that studies of wild primates should not interfere with the primates' day-to-day lives. Increasing the length of field studies opened new lines of inquiry for modern field primatology. Life histories, for example, became the focus of many primatology projects beginning in the 1970s, as researchers wanted to know how animals behaved at each life stage as well as the adaptive value of their behaviors. Any study of life histories requires a project that extends at least as long as the animal's life span, individual recognition of all study subjects, and ideally knowledge of paternity and relatedness between animals.[1] Each of these requirements came with its own methodological challenges, including how to gain hormonal and genetic information while adhering to noninterventionist practices and how to deal with increasingly large amounts of data.

For Western primatologists whose academic and personal obligations limited the time they could spend in the field, long-term field research also presented a host of practical challenges. Who would observe the study subjects and maintain the site when they were researching, teaching, and living in their home countries? How would data and samples be collected and transferred between international field sites and American-based research laboratories? How would observations performed by a number of individuals from different countries and with different levels of educational background be made uniform enough to enable systematic data analysis? The answer to these questions lay in elevating the status and responsibilities of local people from laborers, cooks, and drivers to positions like field managers and researchers. Field sites became permanent or semi-permanent homes for teams of indigenous researchers, some of whom were in charge of observing the study's primate groups day in and day out, while others performed other crucial tasks such as providing researchers with food and transportation.

The development of transnational research teams profoundly changed how primate studies functioned. The methodological and practical implications of creating field teams of permanent indigenous researchers were readily apparent to the primatologists who established long-term field projects in the early 1970s, as were the political ramifications of these scientific projects. Relationships between Western primatologists, indigenous researchers, local communities, and wildlife were always complex and sometimes acrimonious. Western researchers valued primates as study subjects and in terms of conservation, while local people often viewed them as pests. Their differing views of the animals at times collided, especially when access to water and food resources were concerned. Field primate studies frequently took place in areas of Africa with a colonial history, which amplified tensions between local people, wildlife scientists, and park wardens. The loss of Maasai lands and water sources during the construction of Amboseli National Park in Kenya, for example, negatively impacted the relationships between local people and the Western researchers who established long-term field projects in the region.

Employing local people as field researchers met an important practical goal—the sustained collection of observational data—while at the same time it assisted communication and cooperation between local communities and Western primatologists. Despite important contributions to modern primatology, many of the indigenous people were at best unrecognized and at worst depicted quite negatively in primatology publications and popular representations during the 1960s, 1970s, and 1980s.[2] With the rise of third-wave feminism, postcolonial theory, and a growing awareness concerning the intersection of science and social justice beginning in the 1980s, some Western scientists became increasingly uncomfortable with indigenous researchers' invisibility. They wanted to make plain the myriad of practical and intellectual contributions made by indigenous researchers, individuals who frequently lacked formal scientific educations despite having a wealth of knowledge about local flora and fauna. The process of making their contributions visible is ongoing and has included educational opportunities, coauthorship, and attempts by project directors to move away from images of lone Western primatologists in the field and toward depictions of the kinds of international collaboration that now form the heart of modern primatology.

The Amboseli Baboon Project, with a field site in Kenya and laboratories in the United States at Princeton University and Duke University, embodies the methodological and political challenges of modern transnational, long-term fieldwork. In some ways representative of other long-term field projects established by Westerners in the seventies, the Amboseli Baboon Project is noteworthy for having been particularly proactive in recognizing indigenous people's important contributions to modern primatology. Integrating Kenyan researchers has enabled the Amboseli baboons to be continually observed for over forty years, allowing life histories to be constructed for 300 subjects and generating knowledge of the adaptive value of behaviors across generations.

Mathematics and Monkey-Watching:
The Altmanns and the Amboseli Baboon Project

American primatologists Stuart and Jeanne Altmann initially conducted a fourteen-month field study of yellow baboons in what was then called the Masai-Amboseli Game Reserve (later renamed Amboseli National Park) from June 1963 to August 1964. Unlike their contemporaries in such areas as Nairobi National Park, the Amboseli baboons were not accustomed to tourism pressure and subsequent feedings. This was particularly attractive to the Altmanns, especially Stuart, who was keen to avoid the kind of semi-artificial location he had experienced at Cayo Santiago during his doctoral study of rhesus monkeys in the 1950s.

By the end of the 1963–64 Amboseli study, the Altmanns had made nearly 1,500 hours of observations of natural baboon behavior. Like George Schaller's gorilla study, the Altmanns focused on basic, but previously unknown, information regarding how many baboons lived in the Amboseli region, what they ate, where they found water, and when and where they slept. They also examined the migration of males as well as baboons' responses to predators, and they situated the baboons within their broader ecological context by outlining the relationships the baboons had with the various animals that shared their habitat. The Altmanns' observed the baboons from a platform built on top of a Land Rover and dictated what they saw into a recording device. When the Minifon wire recorders they used malfunctioned, they recorded observations by hand.[3]

The Altmanns' fourteen-month study resulted in the book, *Baboon Ecology*, which was written by both of them and published in 1970. Discussion of the role of Africans in their work is completely absent from the text. Instead, the book focuses on describing and quantifying the behaviors of the Amboseli baboons. Unlike earlier texts, *Baboon Ecology* does not include photos of the Western researchers in the field. Instead, it is heavily illustrated with maps, tables, graphs, and photos and pencil sketches of baboons.

Baboons and mathematics took center stage in the pages of *Baboon Ecology*. Keen to quantify their observational data, the Altmanns engaged Stephen S. Wagner, a mathematician affiliated with the Yerkes Regional Primate Research Center, to supplement their own strong proficiencies in mathematics.[4] Throughout *Baboon Ecology*, the Altmanns used mathematics to measure observation bias. For example, when discussing the time of day at which baboons ascend trees to sleep, they pointed out how their data was "biased against late ascents" because at times their observations stopped before the baboons ascended. Probability estimates could be used to minimize or remove such bias, but the Altmanns concluded that in some instances it would be impossible to use mathematics to remove bias from the data sets.[5] Previous primate researchers had not been so open about the role of bias in their observation, and they had not seized the potential of mathematical methodologies to circumvent such bias.

After being in the United States for a few years and spending some months in Kenya in 1969 and again in 1970, the Altmanns decided to return to Amboseli in 1971 to establish a study of baboons that continues today. Although it was not the Altmanns' original intent to establish a multidecade project that became reliant on international relationships and a group of permanent Kenyan researchers, they could not pass up the opportunity to study baboons in such an ideal, if politically complex, field location.

Amboseli is a region in South Kenya, close to the border with Tanzania and just north of Mount Kilimanjaro. The mountain forms the backdrop for Amboseli National Park, which was originally founded in 1948 on lands historically owned by Maasai pastoralists (see figure 18). Forced removal of indigenous peoples is part of the history of many national parks, and the Amboseli Park is no exception. In the late nineteenth century the Maasai had a territory that stretched from northern Kenya to Central Tanzania. By the end of the twentieth century, they had 100,000 square kilometers in northern Tanzania and in southwest Kenya. Thus, for Maasai the establishment of national parks has meant loss of access to essential resources including water supplies.[6]

In 1961, as Kenya moved toward independence, control of the park temporarily shifted to the Kajiado African District Council, which represented local Maasai communities. However, as tourism income increased during the 1960s and 1970s, interruptions in the flow of finances from government officials to local Maasai became common, while growing wildlife populations increased competition for water resources. Maasai pastoralists responded by grazing their cattle within the park boundaries, an area that frequently had relatively rich water and vegetation resources. Meanwhile, Western conservationists and scientists sought to keep cattle out of the park in an attempt to preserve the water and vegetation resources for wildlife. In hopes of bridging these competing motivations, and fueled by divergent views of wildlife and racial overtones of a colonial past, conservationists such as David Western led efforts to establish a community-based approach to conservation in Amboseli. These efforts, which were particularly pursued from 1977, promoted voluntary involvement of local people in conservation and an appreciation on the part of conservationists and wildlife scientists of the costs and benefits that characterize relationships between local people and wildlife.[7]

Amboseli's colonial past continues to shape relationships between Maasai, wildlife, and Western conservationists and field scientists.[8] Tensions

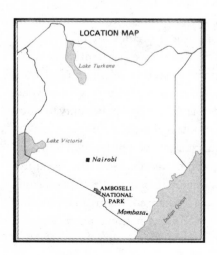

Figure 18. Excerpt of a map of the Amboseli National Park produced by the Kenyan government in 1990. (Map Library, Michigan State University Library)

between these groups have been highest when scientists studied animals—such as the elephant—that can cause significant damage to property and loss of life.[9] The wild savannah baboons (*Papio cynocephalus*) studied by the Amboseli Baboon Project have proven to be a less political beast, with most local people seeing the primate as a pest whose main crime is occasional crop raiding. Nevertheless, establishing a permanent long-term project in 1971 involved many challenges, including finding ongoing funding, coordinating logistics, and adapting to the historical context of the park itself.[10]

Stuart Altmann originally directed the Amboseli Baboon Project. He had received his master's in biology in 1954 under George "Bart" Bartholomew and went on to complete his doctorate at Harvard University under another prominent name in biology, E. O. Wilson. Once at Harvard, in the Department of Medical Zoology, Altmann followed Clarence Ray Carpenter's footsteps by studying howler monkey communication at Barro Colorado in Panama, which was the first of numerous field experiences that spanned Altmann's career.[11] The second such experience was a study of rhesus macaques at Cayo Santiago in Puerto Rico, which was funded by the National Institutes of Health and became the basis of his doctoral thesis.[12]

Unlike her husband, Jeanne Altmann's training began in mathematics. Initially an undergraduate at University of California, Los Angeles, she transferred to Massachusetts Institute of Technology before ultimately graduating with a bachelor's degree in mathematics from the University of Alberta in 1962. While a student at UCLA, she had a summer job at the National Institutes of Health in Maryland, where she met Stuart. Her transfer to MIT coincided with his final year of his PhD thesis. During this time, Jeanne Altmann worked as a computer programmer for Beatrice Whiting, an anthropologist at Harvard. Whiting had conducted research for her six-culture study of child behavior, a project that drew on an international and standardized data set. The experience of being part of such a scientific research project exposed Jeanne Altmann to the demands of international data collection and analysis.

While studying for her various degrees, Jeanne Altmann was also accompanying Stuart in the field, traveling to Amboseli for the first time in 1963 and returning regularly from 1971 onward. After the experience of watching monkeys with her husband, Jeanne Altmann pursued a PhD in behavioral sciences at the University of Chicago within Chicago's Committee on Human Development. After Stuart Altmann stepped down as director, Jeanne Altmann became the primary director of the Amboseli Baboon Project, with Susan Alberts, one of Jeanne Altmann's former graduate students as codirector.[13] Jeanne Altmann's directorship of the Amboseli Baboon Project and her extensive publication record has led to her status as a leader in animal behavior studies.

Jeanne Altmann's 1974 article for *Behaviour*, "Observational Study of Behavior: Sampling Methods," initiated a revolution in the observation of animal behavior. Before 1974, most animal behavior studies had been performed without due consideration about how the method of observation impacted the kinds of data collected. Recognizing this, Altmann summarized several observation techniques, outlining the kinds of questions each method could be used to explore. For example, focal sampling, the observation of one selected individual for a predetermined set of time, was identified as working particularly well for nonsocial behaviors. Her 1974 sampling paper stemmed directly from Altmann's mathematical training, her previous work with the Six Cultures project, and her field experience at Amboseli. Outlining various sampling methods and the consequences of the observational method for data collection, her paper led leaders in the field of animal behavior to question their methodologies and rethink the role of bias in data collection. A central argument of the paper was that the observer's gaze should be turned away from a focus on the easiest to see and most "engaging" animals, commonly males, and toward systematic observation of males, females, and juveniles. The article laid bare bias in observation and specifically bias based on the sex of the study subject.[14]

Susan Alberts shared Jeanne Altmann's interest in issues of gender; indeed, she is an affiliated faculty member of Duke's Women's Studies Program. Her research areas include paternal behaviors and sexual selection, primarily in relation to baboons, although she has also coauthored publications concerning elephant behavior.[15] In collaboration with Jeanne Altmann, Alberts now codirects the Amboseli Baboon Project, a responsibility she likens to having a second "family."[16] This analogy captures the complexity of managing a field study involving finding, maintaining, and coordinating multiyear funding and permits, while also harboring the emotional burden of providing continual employment for local workers who rely on the project for their livelihoods.

Life Histories Occurring in Nature and Synthesized in Databases

From the outset, the Amboseli Baboon Project integrated mathematics and evolutionary biology with a focus on mother-infant relationships and on quantifying the frequency and fitness value of baboon behaviors.[17] Although the work has led to numerous publications, the former research interest is particularly well embodied in Jeanne Altmann's *Baboon Mothers and Infants*, published in 1980, while the latter formed the heart of Stuart Altmann's *Foraging for Survival: Yearling Baboons in Africa* published in 1998.[18] Beginning in the 1970s and increasing thereafter, Stuart and Jeanne Altmann collaborated with mathematicians to test theoretical models with long-term field data.[19] This synthesis of sophisticated mathematics with primate fieldwork enabled the prediction of baboon behavior and quantification of their fitness.[20] Such an advanced application of mathematics distinguished the Altmanns' work from previous studies of primate behavior; indeed zoologist Hans Kummer particularly highlighted the novelty of the quantitative analysis in *Baboon Ecology* in his review of the book for *Science*.[21] Although a significant step forward for primate field studies, the Altmanns' full integration of mathematics into primate research may be seen as an extension of the "statistical frame of mind" that characterized much of biology, ecology, and evolutionary studies during the second half of the twentieth century.[22]

Between 1971 and 1981, a rotating group of Western researchers at Amboseli enabled continuous observation of the behavior of individually identified baboons. One key contributor was Glen Hausfater from the neurobiology and behavior program at Cornell University, who began collecting the life histories of a single baboon group, named Alto's group. Named after its highest ranking female, Alto's group was one of several that formed part of the Amboseli Baboon Project, with each group member receiving a name beginning with the same letter as its mother for recording individual behavioral observations. Hausfater and the project's directors, initially Stuart and Jeanne Altmann and later Jeanne Altmann and Susan Alberts, sought to use data collected from the longitudinal baboon study to compare the fitness of individuals and populations over time. As such, their work contributed to a body of literature referred to as life history theory, which understands an organism's fitness "as a suite with the target of selection being fitness over the lifetime rather than instantaneous fitness or maximization of any single trait."[23] Any life history study relies on making observations of individuals including all births, deaths, and migrations. As Stuart and Jeanne Altmann demonstrated in a 1979 paper, such basic life history data needs to be supplemented by demographic information concerning the age, sex, and kinship of the group and behavioral data concerning social relationships. Dominance and relatedness, for example, play key and interconnected roles in the fitness and life events of a baboon's life history. If a baboon's mother happens to be the highest-ranking female in the group, then that infant will assume a position of social dominance. A baboon's position in the social hierarchy determines access to food, water, and mates and thus significantly affects that individual's fitness. Gaining the kinds of observational data required to compile the life histories required to study fitness in this way demands intense and continual observation over the life spans of numerous individuals.

Making continual observations of baboons requires habituation of the groups to human observers. Therefore, from the start of the Amboseli Baboon Project, Stuart and Jeanne Altmann sought "to maintain a neutral relationship with the baboons, at no time attempting any intervention in anything they did or anything that befell them."[24] For many of the Altmanns' and Albert's study questions, the value of what scientists call "non-interventionist practices" could be upheld. For Jeanne Altmann's mother-infant study, for example, she determined maternity based on which baboon was nursing which infant, thus rendering any genetic testing unnecessary. For other questions, however, it became necessary to track the study subject's hormone levels and DNA. Basic hormonal and DNA data can be obtained by collecting fecal samples in the field, a practice performed by the Kenyan field team without disrupting the baboon's behavior.[25] Here, the main challenge is the transportation of fecal samples across the continents involved in this transnational scientific enterprise. Gaining high quality DNA and RNA data, however, involved brief breaks in the project's non-interventionist approach. Working with fellow primatologist Robert M. Sapolsky, who used darting with tranquilizers to collect blood samples for a study of stress and testosterone in a group of baboons in the Serengeti, Alberts, the Kenyan field team and other members of the project have used darting to collect detailed genetic information on the Amboseli baboons since 2006.[26] These data are used for studies of connections between genotypes and factors impacting a primate's life and death, such as their susceptibility to diseases like malaria. Following Sapolsky's methods—which included darting animals when they were alone and darting only two or fewer animals a day for a maximum of six weeks—enabled the Amboseli researchers to maintain the baboons' habituation to their presence despite the interventionist nature of darting.[27]

The flood of data generated by synthesizing behavioral observations with genetic and hormone data for three hundred baboons compelled Jeanne Altmann and Susan Alberts to develop the BaBase Software Project. The

BaBase database includes each baboon's group membership, social status, sexual cycles, life events, and observations made during focal sampling. The database also holds records concerning weather, dartings, and geospatial locations such as where baboon groups are sleeping and drinking. In the spirit of the democratization of science, this software is downloadable for free from the Amboseli Baboon Project's website.

The Africanization of the Amboseli Baboon Project

In response to the growing challenge of compiling the huge amount of data required to study the life histories of hundreds of baboons and to her growing commitment to issues of social justice, Jeanne Altmann embarked on what Susan Alberts has called the "Africanization" of the Amboseli Baboon Project in 1981. Instead of a rotation of Western researchers, Jeanne Altmann hired Raphael Mututua, a member of a local Maasai community, as a field researcher in 1981. In 1989, the project hired Serah Sayialel as a full-time researcher, now second in command under Mututua, with a third researcher, Kinyua Warutere joining the team in 1995 (see figures 19 and 20).

The central role of the Kenyan field team was, and continues to be, observing the baboons. Since 2000, members of the field team, including some of the more experienced drivers, have been given the additional responsibility of collecting fecal samples. The samples are then transported to laboratories in the United States and used to determine relatedness between group members. The Kenyan field team performs these observations and sample collections year-round, regardless of whether the Western directors are present. Protocols for sample collecting are laid out in the *Guidebook for the Long-Term Monitoring of Amboseli Baboons and Their Habitat: Definitions, Procedures, and Responsibilities,* written by Jeanne and Stuart Altmann in collaboration with Glenn Hausfater in 1981 and later revised by Jeanne Altmann and Susan Alberts. The guidebook is a direct extension of Jeanne Altmann's 1974 article that had demonstrated that observational methods significantly affect data collection and thus the method of observation was of utmost importance in the running of a long-term field study. Ensuring uniformity in data collection was particularly important for the Amboseli Baboon Project, as it came to attract researchers from a range of institutions and backgrounds, pursuing diverse research questions and with varying degrees of field experience and training.

The Kenyan field team has been using the protocol guidebook and standardized sheets for recording baboon behavior on an almost daily basis since

Figure 19. Raphael Mututua, field manager of the Amboseli Baboon Project, holding a small computer for entering behavioral observations. (Photo taken by the author)

Figure 20. Serah Sayialel, assistant field manager, Amboseli Baboon Project. (Photo taken by the author)

1981. Set worksheets and charts were, and still are, provided for all researchers to complete daily, biweekly, or monthly, depending on the behavioral or ecological category. SWERB record sheets, for example, are to be collected all day in an "attempt to obtain, in readily computerized format, information on sleeping grove usage, descent and ascent times, drinking, subgroup formation and the total number of hours that each observer monitored each study group."[28] The S refers to sleeping subgroup, the W to water drinking, the R to range subgroup, and the E and B to the time that observation began and ended. Other worksheets include sexual swelling charts, birth and neonatal assessment sheets, and agonistic interaction matrices. In most cases, observed behaviors are reduced to a letter that represents that behavior.

The SWERB record sheets and the observational goals each represents pose a considerable challenge for the observer. Each morning begins with a census of the group and observation of any sexual skin swelling or reproductive activity. On any given day, all mounting and consortships, all grooming, and all agonistic bouts between individuals also need recording. In addition, the SWERB sheets are updated all day long while the researchers also record all intergroup interactions and all predation on or by the baboons. At least once every two weeks, plants being consumed or otherwise used by the baboons need to be noted along with information concerning what plants are flowering or fruiting. Each month, the development of juveniles is recorded and postnatal condition of newborns observed on day one of life and again within six days. The Kenyan field team has mastered this complex list of tasks. The work sheets they, along with all Western researchers working at the site at any given time, produce now form the bulk of the folders that cover the walls of Jeanne Altmann's Amboseli laboratory at Princeton University with a full set of the records also being housed at Susan Alberts's laboratory at Duke University. Such standardized methods of observation have dominated the project's dataset since 1981, although some longhand notes remain, providing a limited opportunity for expanding beyond the assigned abbreviations.

In some cases, Western researchers trained members of the Kenyan field team to make and record the list of field observations. Jeanne Altmann and Michael Pereira, her graduate student at the time, trained Raphael Mututua in the observation protocols. Stuart Altmann was also involved in Mututua's training, including teaching Mututua the names of plants the baboons ate. Susan Alberts introduced Serah Sayialel to the method of identifying individual baboons, but her comprehensive training came from Mututua and Philip Muruthi, a Kenyan scientist who completed his PhD in the United States and then returned to Africa. Warutere was trained by Muruthi at his own baboon field site, where Muruthi applied the same methods as those used by the Altmanns and Alberts, but his was a study of Anubis baboons. Upon the completion of Muruthi's study, Warutere joined the Amboseli Baboon Project.

A simple apprenticeship model depicting the Kenyan researchers as students of the project's Western directors fails to capture the field team's role, for Mututua, Sayialel, and Warutere are also teachers. The Kenyan field team has the most permanent and in-depth knowledge of each individual baboon in each group, thus they teach Western graduate students who visit the camp how to observe the baboons and how to follow the protocol guidebook. Courtney Fitzpatrick, one of Susan Alberts's graduate students, for example, commented that the Kenyan field team "literally teach you how to *see* the animals." In this context *seeing* means being aware of the animal's "qualities"—such as how the animal moves or its personality—as well as the animal's physical traits.[29] The Kenyan field team also reminds the project directors of the names or "IDs" of certain individuals after they have been away from the field for any significant length of time.

For a long-term study focused on life histories of individual baboons, identification of individual baboons is essential. Although initially trained to key-out the individual baboons according to distinguishing features, each researcher has a moment when he or she simply recognizes the whole animal—the way they walk, their appearance, who they are. Alberts described the process of learning how to identify each individual baboon: "So, this is something that Stuart [Altmann] actually taught me which is when you are first learning the animals you key them out just like you would . . . a bird. . . . Does it have this feature, does it have that feature, if yes then it must be so-and-so, if no then go to this branch. . . . So, at first you key them out and then there is a point at which you start gestalting them and I rarely anymore key them out . . . just like you don't key out people."

Alberts described the moment when you can identify all the baboons in the study groups as "entering a different world," a world she reenters many times as she learns and relearns the IDs of new individuals and of individuals she once knew but has since forgotten. She recollects this process of knowing and then reknowing: "And then every time I came back I would sort of wait for that moment when I could look at every interaction and know who was doing what to whom. And it just brings things into focus in a way that they aren't if you don't know who is doing what to whom. . . . It literally adds another dimension."[30] Echoing Alberts, Mututua recognized a moment when you cease keying out individuals and simply know them like you know individual humans: "From the beginning you keep records. . . . You look at the tails, shapes, . . . coat color, . . . if there is any notch. . . . But as you continue staying with the animals you don't even remember those things—it looks like it's your brother or how you can know a different individual from you."[31]

In addition to the field team's significant contributions to the project in terms of their extensive knowledge of the life histories of individual baboons and systematic data collection, they also play a crucial role in the logistical running of the field site. This includes managing the long-standing relationship between the project and local Maasai. Leasing their land for the field campsite, employing local Maasai in the camp as cooks, drivers, and water collectors, and offering rides to local people when driving around the park maintain relationships with local Maasai. For example, figure 21 shows one of the drivers, Gideon Marinka, who is a member of a local Maasai community. Marinka uses telemetry equipment to locate one of the baboon study groups via the very high frequency (VHF) radio collar worn by one member of each study group (figure 22).

Beyond practical considerations, political and personal relationships between the Amboseli Baboon Project and local Maasai also require regular maintenance. The event in which a number of baboons were killed by local Maasai and then buried in the project's field camp that is discussed in this book's introduction highlights the complex and at times contentious relationship between wildlife scientists and local communities. Mututua and Sayialel were central to efforts to repair this momentary interruption in what had been a close and successful relationship between the Amboseli Baboon Project and local Maasai. They met with local Maasai to reemphasize the positive relationship between the project and local peoples, and local Maasai

Figure 21. Gideon Marinka, a driver for the Amboseli Baboon Project. (Photo taken by the author)

Figure 22. Baboon with VHF collar. (Photo taken by the author)

were employed as "chasers" to chase away any baboons that came too close to livestock in the future. In situations like these, the Kenyan field team operates as diplomats as well as experts in terms of the identities and life histories of the study's subjects.

Mututua described the management of the complex relationship with the local community as the most difficult aspect of his job as field manager, but he also made clear that the relationship between the community and the baboon project was a very positive one: "Actually, after that problem that we had in [March] we have never had [another problem like that]." Furthermore, when asked whether the local community was hostile to the project because of the colonial history of the national park, Mututua stated: "I think the community here actually really like this ... instance of research because they benefit a lot from the research institutions because they create employment that supports their kids, and that really keeps them in a very good relationship with [us]. They don't actually ever think that we are really dominating and we are misusing their land.... We are not pushing them anywhere.... What I believe is they really like us and they really like the project because they are really benefiting from [it]."[32] Certainly, the apparent ability of the Amboseli Baboon Project to maintain a positive connection with local Maasai after the death of the goat and burial of the eleven baboons demonstrated the solid foundation that has been built by Mututua, Sayialel, Warutere, and the project's directors, Alberts and Altmann. In areas where colonial histories inform contemporary relationships between Western scientists, indigenous researchers, and local communities, managing these relationships is essential to maintaining a long-term field project. Such diplomatic skills escape common measures of Western science, such as publications and qualifications, but modern primatology would grind to a stop without them.

Speaking with the Kenyan field team in the summer of 2009 provided me with the opportunity to inquire about their own perspective on their work. Like many of the earlier primate researchers, including Yerkes and zookeeper Belle Benchley, each member of the Kenyan field team emphasized their emotional connection with the primates they studied. Similarly, each researcher expressed that their enjoyment of their work stemmed from their enjoyment of the baboons themselves. Warutere, for example, described the baboons as "social, ... funny, ... and ... also good to work with—no stress.... It's a good job." He continued, "I like it a lot, working here.... If I had to choose between working in an office and working in the group, I would choose to work with the baboons."[33] Sayialel expressed an incredible love for the baboons and her job. When asked how she felt about her research, and specifically her coauthorship of a poster on cannibalism in baboons that is displayed in the field camp's office, she simply said, "I feel that it is something." However, the look on her face communicated a profound feeling of pride that her modesty and cultural background prevented her from expressing verbally. Mututua also emphasized his love for the job, one originally seen as highly unusual by his family and friends. Following baboons "was something very strange because here in Kenya we did not know anything like follow[ing] the animals, ... [writing and recording] the behaviors and things like that.... Even now it is still very hard to explain to people what you do in the field with baboons. But, to me, I became so ... interested in their behavior and when I got to know ... exactly what they are doing I just came to realize they are very smart animals and they are very close to human beings."[34] Warutere's family also initially felt he had chosen an unusual line of employment but now feel such wildlife science and conservation is very important work: "They [his family] think it is funny but at first they thought it was kind of weird work, it was not fun to work with baboons, and they ... laughed at us sometimes. But now they think different[ly]. They think it is a really important job and it's really exciting to work with them because I keep on telling them how funny they are, how good they are to work with."[35] Indeed both Warutere and Sayialel said that their work with the baboons and discussion of this work with their families had led members of their local communities to perceive baboons less as pests, a shift that encouraged their families to advocate for greater tolerance of baboons in their farming communities.

For Alberts, the degree of integration of Kenyan field researchers and employment of local Maasai as cooks and drivers justifies her statement, "This is an *African* camp," indeed, a camp that has been "Africanized" since the 1980s to integrate African workers, researchers, and communities. Jeanne Altmann's attempts to make the indigenous field team's important contributions visible have been central to this process of "Africanization." This process has involved recognizing their contributions in the form of acknowledgements, inclusion in textbooks, graduate opportunities, and coauthorship of posters and articles. When asked for a photo to include in the influential textbook *Primate Societies,* for example, Jeanne Altmann insisted that the editors use a photo of Raphael Mututua, now the field manager, rather than a photo of herself. Other contributors to the project, such as the Amboseli-Longido pastoralist communities who own group ranches around the park and allow the researchers access to their land for observing baboons and lease their land for use as the project's field camp, have also been regularly recognized in Jeanne Altmann's and Susan Alberts's scientific publications.

Providing educational opportunities to Kenyan researchers has also been a central characteristic of the Amboseli Baboon Project since 1984, with the Chicago Zoological Society's Brookfield Zoo providing funds specifically for the training of Kenyan scientists. For example, Philip Muruthi received his bachelor of science degree from the University of Nairobi and went on to study with Dan Rubenstein at Princeton and, with Jeanne Altman serving on his committee, he ultimately received his PhD in 1997.[36] In terms of authorship, Kenyan researchers working with the Amboseli Baboon Project have published as coauthors since 1988, when Muruthi was second author on an article in the *American Journal of Primatology*.[37] Muruthi later returned to Africa and became research director of the African Wildlife Foundation in Nairobi. Mututua and Sayialel coauthored their first paper in 1996 and a book chapter in 2005, which was also coauthored by Warutere.[38]

The Amboseli Baboon Project captures many of the characteristics and challenges of modern primatology, now a science involving long-term international collaborations. It embodies the transition away from spending a few weeks or months in the field, as was done by Clarence Ray Carpenter, and toward studying primates for several decades, as well as the shift from describing basic information and behaviors based on relatively few hours of observation to deep quantitative analysis founded on life histories spanning several generations. The Amboseli Baboon Project also embodies a literal move, as Western researchers began to establish permanent or semi-permanent field studies in the actual locations where their study subjects live.

Local people are no longer employed merely to carry equipment during brief expeditions as they were during the Asiatic Primate Expedition in 1937. With the new longevity of fieldwork and the growing emphasis on life histories, indigenous people have become essential to the intellectual and logistical running of primate field projects. Like Rosalia Abreu, Ben Burbridge and Belle Benchley, local researchers such as Raphael Mututua commonly lack the kind of credentials that lead to the label of "scientist" in Western cultures. Nevertheless, their contributions to data collection and facilitation of long-term field studies in areas with colonial histories were, and are, no less crucial to modern field primatology than the institutional building and methodological developments made by historical actors such as Yerkes and Carpenter during primatology's emergence as a science.

Conclusion

Chimpanzees and humans share 98 percent of their genetic material, but, unlike our primate cousins, humans have created religion, science, art, philosophy, and technology. We have long held up these differences to mark a clean line between humans and other animals. Just five years ago, I could spend hours showing students about the many ways that nonhuman primates are like us—their tool use, their culture, and their emotional lives—only to find students holding tight to all the ways we are apparently unique.

In recent years, however, the similarities between humans and animals have become the increasing focus of conversations in the popular press, academic publications, and scientific articles. Today when I teach courses on the history of primatology and the human-animal boundary, students are eager to share articles from the *New York Times, Huffington Post, National Geographic, Nova,* and *Science* that report how primates have demonstrated their similarities to humans. Tales of primates saving young zoo visitors, grieving the death of a group member, or appearing to keep dogs as pets challenge the boundaries we like to create between us and them. The increasing popularity of such articles, blogs, and viral videos reflect the public's continued fascination with primates' natures and what primates might help us better understand about ourselves.

The wide range of media coverage in blogs, documentaries, magazines, and radio shows provides a modern-day version of primate folklore. However, there is a crucial difference between nineteenth-century and early twentieth-century tales of beasts in the jungle and today's stories about primates. Today, scientists and science, not amateurs and myth, are central to the stories about primates that circulate in the popular press. Remnants of adventure and exotic wilderness locations certainly remain, though, and the stories continue to emphasize solidary Westerners braving untamed jungles. The transnational nature of modern field primatology and the international teams of collaborators that make the science possible are still not represented in popular depictions of primate studies. Instead, like the nineteenth-century myths that preceded them, the stories in the media today typically focus on apes either as mirrors for humans' primeval natures or as examples of animals that appear human-like in various ways.

Sex is—and long has been—at the heart of many primate stories. For example, Daniel Bergner's 2013 book *What Do Women Want? Adventures in the Science of Female Desire* opens with a chapter on animals and has another entire chapter dedicated to just monkeys and rats.[1] In discussing lust in female animals, Bergner shares observations about a rhesus macaque whose kissing and caressing of a prospective mate may be less obvious than the thrusting of a male but is no less indicative of her sexual assertiveness. Like *Life* magazine's 1939 description of the macaques on Cayo Santiago Island who drove at least one male to escape to open water, free-ranging female rhesus monkeys aggressively pursue males for sex. Elaine Blair, who reviewed *What Do Women Want?* for the *New York Times* also weaved together discussion of human and primate sexual desire before asking the age-old question, "Are we that kind of primate?"[2] The main thing that distinguishes Bergner's inquiry from that pursued through the majority of the twentieth century is his focus on female, rather than male, sexuality. After all, when primatology emerged as a scientific field, Yerkes's Committee for Research on Problems of Sex provided funding for the early studies of primate behavior, in turn propelling researchers to examine primate copulation, masturbation, and homosexual or "nonconformist" sexual behaviors. Although the researchers who performed early primate studies often looked at both male and female sexual behaviors, the studies that made their way into the popular press almost always dealt with the primitive, natural, urges of men.

Men's sexual urges and activities have long been naturalized through analogies with animals, and they reached a height in popularity in the 1970s and 1980s when pop sociobiology seeped into publications ranging from *Time* to *Playboy*. Mainly used to provide evolutionary justifications—all too frequently deprived of solid empirical data—for male promiscuity and even rape, animals and particularly primates were used to explore and justify male sexuality into the twenty-first century.[3] Take for example the 1999 self-help book, *The Silverback Gorilla Syndrome: Transforming Primitive Man*, which sought to teach young men to be sensitive by showing them how to control the gorillas within themselves. The assumption that human males have an inner sexual beast continues to go so unquestioned that it forms the basis for even relatively progressive texts promoting healthier sexual relations for the modern man.[4]

Alongside contemporary conversations about how primates shed light on human sexual norms is a growing debate about the ethical consequences of our similarities to other great apes. The emotional connection between field researchers and their animal subjects, so powerfully exemplified by the burial of eleven baboons in the grounds of the Amboseli Baboon Project's field camp, has motivated many field scientists to become animal activists and advocates for conservation. More recently, these ethical concerns have extended to the world of the lab, where primates are commonly used for biomedical experiments. Jane Goodall and the United States Fish and Wildlife Service, for example, have proposed adding captive chimpanzees to the endangered species list. If successful, the two thousand chimpanzees currently in captivity in the United States will need to be retired as experimental subjects. The National Institutes of Health has also declared that it will be retiring 300 of their 360 chimpanzee subjects.[5] Recent calls for primate protections are the product of years of campaigning by prominent animal advocates such as the philosopher Peter Singer, the ecologist and evolutionary biologist Marc Bekoff, and interdisciplinary groups such as the Great Ape Project, which has advocated for a World Declaration on Great Apes to protect them from death, torture, and inhumane captive conditions since 1994.[6]

Concerns for primates' welfare do not simply reflect the attitudes of a small group of scientists and animal advocates but rather a broader sea change in Americans' view of primates. Films such as *Project Nim*, which chronicled the life of a chimp who was the subject of language studies at Columbia University, and *One Small Step* about the primates used in the American space program, emerged during this time of growing popular concern for animal welfare.[7] In addition, the Smithsonian Institution, graphic novels, and National Public Radio have all recently examined studies of nonhuman primates in connection to human behavior. For example, when the radio show *This American Life* explored attachment and adoption, it chose Harry Harlow's research of rhesus macaques as the starting point for a discussion of the important role played by human contact in the development of our ability to feel and express love.[8] Whether in communication research, space exploration, or studies of human love and attachment, the similarity between humans and other primates has made our primate cousins a very valuable experimental subject. At the same time, the humanness captured in the primates' expressions of fear and pain inspires our compassion toward them and at times horror for what they endured in the name of science.

Western primatologists and indigenous researchers who have spent prolonged time observing primates in the wild cannot help but be struck by the similarities between humans and other primates. Their commonalities serve practical ends, such as helping justify research projects in grant proposals and publications, but they also operate at a personal level, sculpting the relationships that form between scientists and their animal subjects. During the first half of the twentieth century, both Yerkes, with his strict adherence to scientific standards, and Belle Benchley, who operated without formal scientific training, openly expressed the affection they felt toward the primates they studied, housed, and embraced. Later in the twentieth century, Jane Goodall and Dian Fossey professed the love they felt for their fellow primates, and both women became iconic symbols of primate welfare and conservation. Despite a long history of declaring attachment with their primate subjects—or perhaps because of it—contemporary American primatologists continue to struggle with the extent to which they ought to express how they feel about their primate subjects, animals they had observed for years rather than merely weeks or months. At the Amboseli Baboon Project, all of the researchers—American and Kenyan alike—obviously deeply care for the animals whose life histories they have observed, described, quantified, and analyzed for decades. In many ways, the colleagues and primate participants of long-term studies like Amboseli Baboon Project become extensions of researchers' own family and friend networks.

Primatologists' recognition of nonhuman primates as participants in research may be the latest step in a series of efforts to be more inclusionary. Like the discipline of history, which has made a conscious effort to integrate previously ignored groups, primatology has gradually come to accept the importance of females, both in terms of the roles they play in primate societies and the important contributions of female scientists. During the 1960s, some prominent male researchers failed to look beyond male primates when they sought to explain how social groups functioned. The role of female primates initially appeared limited to that of mother and mate. In the 1970s, when women primatologists began entering the field, the multifaceted roles of female primates came to the foreground. Primatologists and social scientists have offered competing interpretations of this shift.[9] Some have argued that, once women primatologists entered the field, they were naturally drawn to exploring the activities of the primates of their own sex. Others have claimed that it was not until 1974, when Jeanne Altmann published her revolutionary sampling paper, that researchers even had the methods necessary to distribute their observations across males, females, and juveniles. Certainly, when I was first in the field, the larger males caught my eye and I was left believing that without robust sampling methods a primatologist would make many more observations of adult males than females or juveniles. We can see that the introduction of women to the discipline helped—one way or another—primatologists better understand primates.

The ultimate impact of the increasing number of women in primatology can certainly be debated, but primatology has unquestionably attracted and retained a relatively high number of female researchers as early as the 1920s and particularly from the 1970s onward. Mentorship, media coverage, and openness towards integrating family life with long stints in the field allowed women to play a more prominent role in primatology than they had played in most other natural sciences. The Amboseli Baboon Project's field camp, for example, includes a large wooden play structure by the researchers' tents and a swing outside of the laboratory, which allows the researchers to keep an eye on their children while they test field samples (see figure 23). The integration of women into primatology has led scientists and humanists alike to ask whether the field represents what primatologist Linda Marie Fedigan has called the "goddess discipline."[10]

Primatologists' recognition of indigenous researchers' important contributions to primate studies came later and more gradually than did the integration of Western women. Coverage of primatology in the popular press continues to focus on white Western researchers, frequently women, who appear to be alone in the wilderness while they study wild primates. This trope has been so persistent in part due to the public's continued perception of Africans as poachers rather than conservationists or scientists. For example, a 2012 article in *National Geographic* included a photograph of two Africans standing in front of a dead elephant with its tusks removed. Below the photo, small print identified the Africans as park wardens. Nevertheless, when I showed my students the image, every one of them identified the Africans as poachers. The American public is simply not used to seeing images of Africans as protectors of wildlife or as animal advocates. In particular,

Figure 23. Play structure by researcher's tent in Amboseli Baboon Project Field Camp. (Photo by the author)

African women go unrepresented—appearing neither as poacher nor as protector—despite being critical to the success of long-term field studies like the Amboseli Baboon Project and Amboseli Elephant Project. Jeanne Altmann and Susan Alberts have led the way in including photos of African researchers in their textbooks, journal articles, and websites, but mainstream popular culture has been slow to more thoughtfully integrate images and references to the positive contributions made by Africans to the preservation of their own flora and fauna.

The integration of African researchers into long-term field projects was in part motivated by issues of social justice and inclusion, but it was also a response to the practical challenges of collecting large amounts of data for primate life histories. Field primatology is now a transnational enterprise that spans both decades and continents, creating methodological challenges including how to manage the ever-accumulating data. Characteristic of many modern sciences, the "data deluge" motivated a small group of primatologists, including Susan Alberts and Karen Strier from the University of Wisconsin, to develop the Primate Life History Database (PLHD) in 2007.[11] The PLHD houses demographic data from seven long-term field projects, each exceeding twenty-five years in duration, with the goal of integrating data and making it more accessible to members of the public and scientific community alike. Like the Amboseli Baboon Project, the PLHD involves combining and sharing data and thus requires standardization. The scientists behind the PLHD, however, are responding to these challenges on a new level, bringing together life histories from multiple projects rather than one.

Contemporary long-term field primatology studies are also pushing the boundaries of traditional science funding structures, which have failed to keep up with the increasing duration of fieldwork. Earning a doctorate in primatology now involves eighteen months or more of continuous field observations before a graduate student can even begin analyzing data and start to write a dissertation. Field projects may last for decades, but grants continue to be limited to at most three to five years, which forces directors and students to continually apply for grant renewals. The international nature of primatology requires directors to also continually apply for research permits and thus navigate constantly changing political waters, often with the aid of the indigenous researchers they employ. Solutions to the fundamental challenges to long-term field primatology, and indeed long-term field science in general, remain to be found.

Studying the emergence and operation of modern primatology has taken me on a scholarly and a personal journey. My experience of watching baboons in the wild left me with a heightened awareness of our similarity to other primates, even those outside the classification of the great apes. When I went through labor and lactation, my baby and I exhibited many traits I saw among the nonhuman primates in the field. There is no doubt that we are truly part of the animal kingdom. For me, that claim was no longer one merely grounded in academic scholarship; it was a literal reality. This realization parallels that of primate researchers in the field. If you watch primates long enough, you cannot fail to realize they are a lot like us and we are a lot like them. As a historian of primatology, I have come to understand I am a truly a primate observing primates observing primates.

Notes

Abbreviations

CRC C. Ray Carpenter Papers, Pennsylvania State University Library

RDS Richard D. Sparks Papers, Series I, University of Arizona Archive

RMY Robert Mearns Yerkes Papers, Manuscripts and Archives, Yale University Library

Introduction

1. For a detailed analysis of local views of baboons and other wildlife in Amboseli, see Maria Joana Ferreira de Roque de Pinho, "'Staying Together: People-Wildlife Relationships in a Pastoral Society in Transition, Amboseli Ecosystem, Southern Kenya" (PhD dissertation, Colorado State University, 2009).

2. Interview with Serah Sayialel, Amboseli, June 24, 2009.

3. Personal communication between Alberts and Montgomery, September 15, 2010.

4. I use the word *local* to represent individuals who live near Amboseli and the terms *indigenous* or *Kenyan researcher* to identify researchers who are from Kenya but not necessarily the local area. As Christine Walley has discussed, such terms are not ideal. See Christine Walley, *Rough Waters: Nature and Development in an East African Marine Park* (Princeton, NJ: Princeton University Press, 2004).

5. Sally Gregory Kohlsedt, *The Formation of the American Scientific Community: The American Association for the Advancement of Science, 1848–60* (Urbana: University of Illinois Press, 1976); and Nathan Reingold, ed., *Science in America since 1820* (New York: Science History Publications, 1976).

6. Some examples of the literature examining professionalization of the life sciences include Adrian Desmond, "Redefining the X Axis: 'Professionals,' 'Amateurs' and the Making of Mid-Victorian Biology—A Progress Report," *Journal of the History of Biology* 34 (2001): 3–50; Samuel Alberti, "Amateurs and Professionals in One County: Biology and Natural History in Late Victorian Yorkshire," *Journal of the History of Biology* 34 (2001): 115–47; Mark Barrow, "The Specimen Dealer: Entrepreneurial Natural History in America's Gilded Age," *Journal of the History of Biology* 33 (2000): 493–534; and N. Jardine, J. A. Secord, and E. C. Spary, *Cultures of Natural History* (Cambridge: Cambridge University Press, 1996).

7. Stuart Altmann, ed., *Social Communication among Primates* (Chicago: Chicago University Press, 1967), 376, and Harold Coolidge to Carpenter, September 21, 1970, in Clarence Ray Carpenter's Papers, Penn State University Archives, Box 7, Folder 38.

8. The role of indigenous peoples in an amateur's study of primate vocalizations in

the nineteenth century is nicely discussed in Jeremy Rich, *Missing Links: The African and American Worlds of R.L. Garner, Primate Collector* (Athens: University of Georgia Press, 2012). An important source when analyzing the role of race in science continues to be Sandra Harding, *The "Racial" Economy of Science* (Bloomington: Indiana University Press, 1993).

9. Bruno J. Strasser, "The Experimenter's Museum: GeoBank, Natural History, and the Moral Economies of Biomedicine," *ISIS* 102 (2011): 60–96; and Robert Kohler, *Lords of the Fly: Drosophila Genetics and the Experimental Life* (Chicago: University of Chicago Press, 1994).

10. Jean Donnison, *Midwives and Medical Men: A History of the Struggle for the Control of Childbirth* (London: Historical Publications, 1988).

11. Pamela J. Asquith has published extensively on the history of Japanese primatology. See, for example, Pamela J. Asquith, "Japanese Science and Western Hegemonies: Primatology and the Limits Set to Questions," in *Naked Science: Anthropological Inquiry into Boundaries, Power and Knowledge,* ed. Laura Nader (New York: Routledge, 1996); Pamela J. Asquith, "Anthropomorphism and the Japanese and Western Traditions in Primatology" in *Primate Ontogeny, Cognition and Social Behavior,* ed. Phyllis C. Lee and James G. Else (New York: Cambridge University Press, 1986); and Pamela J. Asquith, "The Inevitability and Utility of Anthropomorphism in Description of Primate Behavior," in *The Meaning of Primate Signals,* ed. Rom Harre and Vernon Reynolds (New York: Cambridge University Press, 1984), 138–76.

1. Separating Fact from Fiction

1. Londa Schiebinger, *Nature's Body: Gender in the Making of Modern Science* (New Brunswick, NJ: Rutgers University Press, 1993), 79.

2. Edward Tyson, *Orang-Outang, sive Homo Sylvestris: Or, the Anatomy of a Pygmie* (1699; 1966 Facsimile with introduction by Ashley Montagu; London: Dawsons of Pall Mall), 92–94.

3. Andreas Vesalius, *On the Fabric of the Human Body.* 1543. (Reprint, San Francisco: Normal Pub, 1998). Emelin Miller contextualizes Tyson's work within traditions of natural history and medical anatomy in "Making Natural History Anatomical: Edward Tyson and Comparative Anatomy in Seventeenth-Century England" (Presentation, History of Science Society Annual Meeting, Chicago, IL, November 8, 2014.)

4. Harriet Ritvo, *The Platypus and the Mermaid and Other Figments of the Classifying Imagination* (Cambridge, MA: Harvard University Press, 1997), 92.

5. Clare Simmons, "A Man of Few Words: The Romantic Orang-Outang and Scott's 'Count Robert of Paris,'" *Scottish Literary Journal* 17:1 (1990): 21–34.

6. Alfred Crowquill, "Mr. Chimpanzee, the Disappointed Traveller," *Bentley's Miscellany* 8 (1840): 490–93.

7. Charles Darwin, *Descent of Man, and Selection in Relation to Sex* (1871; Princeton: Princeton University Press, 1981), 105.

8. For more on Darwin's use of anthropomorphism in *The Expression of Emotions in Man and Animals,* see Eileen Crist, *Images of Animals: Anthropomorphism and Animal Mind* (Philadelphia: Temple University Press, 1999).

9. The amount of literature within the interdisciplinary field of animal studies on

this discomfort is immense. A good starting point is Erica Fudge, *Animal* (London: Reaktion Books, 2002).

10. Janet Browne, "Constructing Darwinism in Literary Culture," in *Unmapped Countries: Biological Visions in Nineteenth-Century Literature and Culture,* ed. Anna Julia (London: Anthem Press, 2005), 55–69; Julia Voss, "Monkeys, Apes and Evolutionary Theory: From Human Descent to King Kong," in *Endless Forms: Charles Darwin, Natural Science and the Visual Arts,* ed. Diana Donald and Jane Munro (New Haven: Yale University Press, 2009), 215–37; J. E. Jones, "Simians, Negroes, and the 'Missing Link': Evolutionary Discourses and Transatlantic Debates on 'The Negro Question,'" in *Darwin in Atlantic Cultures: Evolutionary Visions of Race, Gender and Sexuality,* ed. J. E. Jones and P. B Sharp (New York: Routledge, 2010), 191–207; and S. D. Bernstein, "Ape Anxiety: Sensation Fiction, Evolution, and the Genre Question," *Journal of Victorian Culture* 6, 2: 250–71.

11. Janet Browne and S. Messenger use this term in "Victorian Spectacle: Julia Pastrana, the Bearded and Hairy Female," *Endeavour* 27, 4 (2003): 155–59.

12. Lindsey B. Churchill, "'What Is It?' Difference, Darwin and the Victorian Freak Show," *Darwin in Atlantic Cultures: Evolutionary Visions of Race, Gender and Sexuality,* ed. in J. E. Jones and P. B. Sharp (New York: Routledge, 2010), 203.

13. Browne and Messenger, "Victorian Spectacle," 155–59. See also Kimberly Hamlin, "The 'Case of a Bearded Woman': Hypertrichosis and the Construction of Gender in the Age of Darwin," *American Quarterly* 63, 4 (2011): 955–81. For more on the history of freak shows, see Pascal Blanchard, Gilles Boëtsch, and Nanette Jacomijn Snoep, eds., *Human Zoos: The Invention of the Savage* (Paris: Musée du Quai Branly, 2011); Lillian Craton, *The Victorian Freak Show: The Significance of Disability and Physical Differences in Nineteenth-Century Fiction* (New York: Cambria Press, 2009); and Robert Bogdan, *Freak Show: Presenting Human Oddities for Amusement and Profit* (Chicago: University of Chicago Press, 1988).

14. Voss, "Monkeys, Apes and Evolutionary Theory," 216

15. Paul B. Du Chaillu, *Explorations and Adventures in Equatorial Africa* (New York: Harper & Brothers, 1862), 322. Du Chaillu is also the subject of a recent popular book: Monte Reel, *Between Man and Beast: An Unlikely Explorer and the African Adventure That Took the Victorian World by Storm* (New York: Anchor, 2013).

16. Kelly Enright, *The Maximum of Wilderness: The Jungle in the American Imagination* (Charlottesville: University of Virginia Press, 2012), 44–45.

17. *King Kong,* Produced by Merian C. Cooper and Ernest B. Schoedsack (1933; Burbank, CA: Warner Home Video, 1999), DVD.

18. Voss, "Monkeys, Apes and Evolutionary Theory," 215–37.

19. Fred D. Pfening Jr. and Richard J. Reynolds III, "The Ringling-Barnum Gorillas and Their Cages" *Bandwagon* November–December 2006: 4–29.

20. Roger Conant, "Gorilla Round-Up: Concerning Each of the Sixteen Gorillas Living in these United States," *Scientific American* June 1943: 246–48.

21. Richard and Sally Glendinning, *Gargantua: The Mighty Gorilla* (New York: Reader's Digest Services, 1977).

22. Anonymous, *An Essay Towards the Character of the Late Chimpanzee, who died Feb. 23, 1738–9* (London: L. Gilliver and J. Clarke, 1739), 8.

23. C. Lloyd Morgan, *Animal Sketches* (London: E. Arnold, 1891), 65.

24. For more on Morgan's Canon, see Donald A. Dewsbury, *Comparative Psychology in the Twentieth Century* (Stroudsburg, PA: Hutchinson Ross, 1984); and Gregory Radick, "Morgan's Canon, Garner's Phonograph, and the Evolutionary Origins of Language and Reason," *British Journal for the History of Science*, 33: 3–23.

25. Gregory Radick, *The Simian Tongue: The Long Debate about Animal Language* (Chicago: University of Chicago Press, 2007), 117.

26. For more on Garner, see Jeremy Rich, *Missing Links: The African and American Worlds of R. L. Garner, Primate Collector* (Athens: Georgia University Press, 2012); and Radick, *The Simian Tongue*.

27. Anonymous, *Joe the Chimpanzee and Other Stories* (New York: Werner, 1899), 19.

28. Ibid., 16.

29. Edgar Rice Burroughs, *Tarzan and the Apes* (London: Methuen & Co., 1917), 254. For more on Tarzan, see Alex Vernon, *On Tarzan* (Athens: University of Georgia Press, 2008).

30. See John Sorenson, *Ape* (London: Reaktion Press, 2009), 25–27.

31. Alfred Russel Wallace, *The Malay Archipelago: The Land of the Orang-Utan and the Bird of Paradise* (London: Macmillan and Co, 1906), 39.

32. Ibid., 32–34.

33. William Atherton Dupuy, *Our Animal Friends and Foes* (Chicago: John C. Winston, 1940), 181.

34. E.G. Boulenger, *Wild Life the World Over* (New York: Wise, 1947), 95.

35. *King Kong*, Directed and Produced by Peter Jackson (2005).

36. Edward Tyson, "A Philological Essay Concerning the Pygmies of the Ancients," in *Orang-Outang, sive Homo Sylvestris*, 1.

37. Voss, "Monkeys, Apes and Evolutionary Theory," 215–37.

38. Anonymous, "With Mr. Gorilla's Compliments," *Temple Bar* 3 (1861): 482.

39. Ibid., 491.

40. For more on tea parties and zoos and in particular how the practice connects with using humor to express our discomfort with the humanness of chimpanzees, see David Hancocks, "Zoo Animals as Entertainment Exhibitions," in *Cultural History of Animals*, ed. Linda Kalof (New York: Berg, 2007).

2. Venturing out of the Lab and into the Wild

1. For biographical information, see George M. Haslerud, "Introduction," in *The Mental Life of Monkeys and Apes* by Robert M. Yerkes (1916; New York: Scholar's Facsimiles and Reprints, 1979), v–ix; and a short entry in *Concise Dictionary of Scientific Biography* (New York: Charles Scribner's Sons, 1981), 743. See also Robert M. Yerkes, "Robert Mearns Yerkes: Psychobiologist," in *History of Psychology in Autobiography*, vol. 2, ed. Carl Murchison (Worcester, MA: Clark University Press, 1932), 381–407. For a history of the laboratories, see Donald A. Dewsbury, *Monkey Farm: A History of the Yerkes Laboratories of Primate Biology, Orange Park, Florida 1930–1965* (Lewisburg, PA: Bucknell University Press, 2005). For works focused on Yerkes intelligence testing and eugenics, see James Reed, "Robert M. Yerkes and the Mental Testing Movement,"

in *Psychological Testing and American Society*, ed. Michael M. Sokal (New Brunswick, NJ: Rutgers University Press, 1987), 75–94; Wade Pickren, "Robert Yerkes, Calvin Stone, and the Beginning of Programmatic Sex Research by Psychologists, 1921–1930," *American Journal of Psychology* 110, no. 4 (1997): 603–19; Hamilton Cravens, "Scientific Racism in Modern America, 1870s–1990s," *Prospects* 21 (1996): 471–90; and James Reed, "Robert M. Yerkes and the Comparative Method," in *Historical Perspectives and the International Status of Comparative Psychology*, ed. Ethel Tobach (Hillsdale, NJ: Erlbaum, 1987), 91–101.

2. For a discussion of the subjects of Yerkes' research, including its contribution to psychology, see John B. Watson, *An Introduction to Comparative Psychology* (New York: Holt, Rinehart and Winston, 1967).

3. Robert M. Yerkes, *An Introduction to Psychology* (New York: Henry Holt and Company, 1911).

4. See, for example, Pickren, "Robert Yerkes, Calvin Stone," 603–19.

5. Robert M. Yerkes, "The Mental Life of Monkeys and Apes: A Study of Ideational Behavior," *Behavior Monographs* 3, no. 1 (1916), 8.

6. The Yale Laboratories of Primate Biology were renamed the Yerkes Laboratories upon Yerkes' retirement from the Laboratories in 1941; he retired from Yale in 1944. He continued to be directly and indirectly involved in primate research until shortly before his death in 1956.

7. Adele E. Clarke, *Disciplining Reproduction: Modernity, American Life Sciences, and "The Problem of Sex"* (Berkeley: University of California Press, 1998); and Pickren, "Robert Yerkes, Calvin Stone," 603–19.

8. Helen M. Pyciour, Nancy G. Slack, and Pnina G. Abir-Am, eds., *Creative Couples in the Sciences* (New Brunswick, NJ: Rutgers University Press, 1996); and A. Lykknes, D.L. Optiz, and B. Vann Tiggelen, *For Better or for Worse? Collaborative Couples in the Sciences* (Basel, Switzerland: Springer, 2012).

9. Robert M. Yerkes and Ada Yerkes, *The Great Apes, A Study of Anthropoid Life* (New Haven: Yale University Press, 1929).

10. Richard W. Dukelow, *The Alpha Males: An Early History of the Regional Primate Research Centers* (New York: University Press of America, 1995), 112.

11. Robert M. Yerkes and Ada Yerkes, "Individuality, Temperament, and Genius in Animals," *American Museum Journal* 17 (1917): 234–43; Robert M. Yerkes and Ada Yerkes, "The Comparative Psychopathology of Infrahuman Primates," in *The Problems of Mental Disorder, A Study Undertaken by the Committee of Psychiatric Investigations, National Research Council*, ed. Madison Bentley (New York: McGraw-Hill, 1934), 327–38; Robert M. Yerkes and Ada Yerkes, "Social Behavior in Infrahuman Primates," in *A Handbook of Social Psychology*, ed. Carl Murchison (London: Oxford University Press, 1935), 973–1033; and Robert M. Yerkes and Ada Yerkes, "Nature and Conditions of Avoidance (Fear) Response in Chimpanzee," *Journal of Comparative Psychology* 21 (1936): 53–66.

12. Robert M. Yerkes and David Yerkes, "Concerning Memory in the Chimpanzee," *Journal of Comparative Psychology* 8 (1928): 237–71.

13. For information concerning the kinds of questions examined by Yerkes, especially in relation to social control and organization, see Donna Haraway, *Primate*

Visions: Gender, Race, and Nature in the World of Modern Science (New York: Routledge, 1989), 59–83.

14. It was this element of primate research that contributed to Yerkes's difficulties in gaining promotions and support at Harvard; see Reed, "Robert M. Yerkes and the Mental Testing Movement," 81.

15. Yerkes, "Creating a Chimpanzee Community," *Yale Journal of Biology and Medicine* 36, no. 3 (1963): 219.

16. Ibid., 219–20.

17. I thank Helen Veit for pointing out the connection between how Yerkes discussed housekeeping and how domestic tasks were commonly discussed during this period.

18. Richard Sparks to Robert Yerkes, November 21, 1931. RDS, Folder 11.

19. Yerkes, "Creating a Chimpanzee Community," 214.

20. Yale Anthropoids to Rockefeller Philanthropoids, May 27, 1929, Rockefeller Foundation to Yale Anthropoids, May 31, 1929. RMY, Box 57, Folder 1089.

21. Robert M. Yerkes, "The Harvard Laboratory of Animal Psychology and the Franklin Field Station," *Journal of Animal Behavior* 4, no. 3 (1914): 181.

22. Ibid., 182.

23. Ibid.

24. Phillip J. Pauly, "Summer Resort and Scientific Discipline: Woods Hole and the Structure of American Biology, 1882–1925," in *The American Development of Biology*, ed. Ronald Rainger, Keith R. Benson, and Jane Maienschein (New Brunswick, NJ: Rutgers University Press, 1991), 121–50.

25. See, for example, Donald L. Opitz, "This House Is a Temple of Research:" Country-House Centers For Late Victorian Science" in *Repositioning Victorian Sciences: Shifting Centers in Nineteenth Century Scientific Thinking*, ed. David Clifford, Elizabeth Wadge, Alex Warwick, and Martin Willis (London: Anthem Press, 2006), 143–241.

26. Robert M. Yerkes and Blanche W. Learned, *Chimpanzee Intelligence and its Vocal Expressions* (Baltimore: Williams and Wilkins, 1925), 37.

27. T. Everett Harré, "A Mansion for Monkeys: A Visit with One of the Strangest Women of Our Time," *Hearst's International Cosmopolitan*, April 1930. RMY, Box 1, Folder 9.

28. Robert M. Yerkes, *Almost Human* (New York: The Century Co., 1925), esp. chapter 1.

29. Personal communication with Eric del Junco, great-grandson of Rosalia Abreu.

30. "Senora Abreu Dead: A Cuban Patriot," *New York Times*, November 5, 1930, 21.

31. William Morton Wheeler to Yerkes, November 1, 1916. RMY, Box 51, Folder 999.

32. For more on Cunningham and her gorilla, John Daniel, see Alyse Cunningham, "A Gorilla's Life in Civilization," *Zoological Society Bulletin* 24 (September 1921): 118–24; and Carole Jahme, *Beauty and the Beasts* (New York, Soho Press, 2003), 17–18. Nadia Kohts conducted studies of primate behavior at the Zoopyschologiches Laboratoriam in Moscow.

33. Based on the extensive use of Abreu's observations, it seems somewhat problematic that she did not receive coauthorship of the work. However, she was heavily

thanked in the acknowledgments. This and Yerkes's encouragement of Blanche W. Learned to take first authorship of their book, *Chimpanzee Intelligence and Its Vocal Expression* (1925), an offer she declined, points to his recognition of female colleagues. Yerkes to Learned, April 8, 1924, Learned to Yerkes, April 13, 1924. RMY, Box 36, Folder 693.

34. Yerkes, *Almost Human,* esp. chapters 11 and 12.

35. Richard Sparks, "Congo: A Personality," *Field and Stream*, 1926: 19–20

36. Mark V. Barrow Jr., *A Passion for Birds: American Ornithology after Audubon* (Princeton: Princeton University Press, 1998), 30–31; Ben Burbridge, "The Gorilla Hunt," *Forest and Stream* 96, no. 11 (1926): 645–49, 680; Ben Burbridge, "The Gorilla Hunt: Part II," *Forest and Stream* 96, no. 12 (1926): 712–15, 744; Ben Burbridge, "The Gorilla Hunt: Part III, The Capture of Bula Matadi," *Forest and Stream* 97, no. 1 (1927): 5–7, 45; Ben Burbridge, "The Gorilla Hunt: Part IV, The Tragedy on Mount Mikeno," *Forest and Stream* 97, no. 2 (1927): 70–71, 104; and Ben Burbridge, "The Gorilla Hunt: Part V, Conclusion," *Forest and Stream* 97, no. 3 (1927): 136–37.

37. Burbridge, "The Gorilla Hunt," 645.

38. For a discussion of the curiosity of gorillas and how Burbridge used this to capture their behavior on film, see Burbridge, "The Gorilla Hunt," 680.

39. Burbridge, "Gorilla Hunt: Part V," 137.

40. Ben Burbridge, *Gorilla: Tracking and Capturing the Ape-Man of Africa* (London: George G. Harrap and Company, 1928), 255–59.

41. For a more detailed account of Congo's arrival in the United States and her biography, see Georgina M. Montgomery, "'Infinite Loneliness': The Life and Times of Miss Congo," *Endeavour* 33, no. 3 (2009): 101–5. Congo and other gorillas displayed by The Ringling-Barnum Circus is discussed in Fred D. Pfening Jr. and Richard J. Reynolds III, "The Ringling-Barnum Gorillas and Their Cages," *Bandwagon*, November–December 2006: 4–29.

42. For a description of the techniques used by Burbridge, see Burbridge, "The Gorilla Hunt," 649 and 680.

43. Burbridge, "The Gorilla Hunt: Part II," 712.

44. Robert M. Yerkes, "The Mind of a Gorilla," *Genetic Psychology Monographs* 2, nos. 1 and 2 (1927): 27.

45. See Yerkes, "The Mind of a Gorilla;" Robert M. Yerkes, "The Mind of a Gorilla: Part II, Mental Development," *Genetic Psychology Monographs* 2, no. 6 (1927): 377–551; and RMY, Box 131, Folder 2238.

46. Yerkes, "Mind of a Gorilla, Part II," 483–84.

47. Yerkes, "The Mind of a Gorilla," 121.

48. Richard Sparks to Harold Bingham, February 5, 1932. RDS, Folder 4.

49. Richard W. Burkhardt Jr. has discussed the emphasis placed on behavioral development by American comparative psychologists in contrast with European ethologists who dealt more with questions of evolution and function. See Richard W. Burkhardt Jr., *Patterns of Behavior: Konrad Lorenz, Niko Tinbergen, and the Founding of Ethology* (Chicago: University of Chicago Press, 2005), 384.

50. Yerkes, "The Mind of a Gorilla," 28.

51. Yerkes pointed out that the mountain gorilla was not studied in any detail until 1920, with some information being collected by Carl Akeley and Mary Hastings Bradley. See Carl E. Akeley, *In Brightest Africa* (New York: Garden City, 1923); and Mary Hastings Bradley, *On the Gorilla Trail* (New York: D. Appleton, 1922).

52. Yerkes, "The Mind of a Gorilla," 7; Sparks also wrote an article on Congo and her value to science; see Sparks, "Congo a Personality," 18–20, 72.

53. Yerkes, "The Mind of a Gorilla," 8–9.

54. Yerkes, "The Mind of a Gorilla: Part III, Memory," *Comparative Psychology Monographs* 5, no. 2 (1928): 61.

55. Yerkes, "The Mind of a Gorilla," 32.

56. Ibid., 34.

57. Yerkes, "The Mind of a Gorilla, Part II," 520.

58. Yerkes, "The Mind of a Gorilla," 148

59. Ibid., 150–51.

60. Ibid., 154.

61. Ibid., 153.

62. For more on the Laura Spelman Rockefeller Memorial Fund, see Reed, "Robert M. Yerkes and the Mental Testing Movement," 86–87.

63. RMY, Box 131, Folder 2238.

64. Yerkes, "The Mind of a Gorilla: Part III," 1–2.

65. Ibid., 7.

66. Ibid., 14.

67. Yerkes frequently used the term "supplement" when discussing the relationship between lab and field studies. See, for example, Robert M. Yerkes, "Foreword," *Carnegie Institution of Washington Publication* 426 (1931): 2. Yerkes mentioned his views of the field in many of his publications. See, for example, Robert M. Yerkes, "The American Society of Naturalists;" and Robert M. Yerkes, "Yale Laboratories of Comparative Psychobiology," *Comparative Psychology Monographs* 8, no. 3 (1932): 1–31. See also Haraway, *Primate Visions*, 75–76. Richard Burkhardt has discussed Yerkes' value of the field while favoring laboratory methods for his own work. See Richard W. Burkhardt Jr., "The *Journal of Animal Behavior* and the Early History of Animal Behavior Studies in America," *Journal of Comparative Psychology*, vol. 101, no. 3 (1987): 223–30.

68. Yerkes, "Yale Laboratories of Comparative Psychobiology," 10.

69. Yerkes, "Foreword," iv–v.

70. Robert M. Yerkes, *Chimpanzee: A Laboratory Colony* (New Haven: Yale University Press, 1943), 40.

71. Ibid., 48.

72. See Harold C. Bingham, "Gorillas in a Native Habitat," *Carnegie Institution of Washington Publication* 426 (1931): 3–66; and "Gorilla Changes Nest Each Night, Studies Reveal," *New York Herald Tribune*, March 16, 1930. RMY, Box 58, Folder 1109.

73. John Merriam to Yerkes, February 9, 1929. RMY, Box 58, Folder 1109.

74. Bingham, "Gorillas in a Native Habitat," 4.

75. See, for example, Mary Jobe Akeley, "Africa's Great National Park: Carl Akeley's Gorilla Sanctuary and Biological Survey Station Becomes a Reality: The Formal Inauguration of the Parc National Albert at Brussels," *Natural History* 29, no. 6 (1929):

638–50. For a history of the creation of the gorilla sanctuary, see Jeannette Eileen Jones, "'Gorilla Trails in Paradise:' Carl Akeley, Mary Bradley, and the American Search for the Missing Link," *The Journal of American Culture* 29, vol. 3 (2006): 321–36.

76. Despite the formidable scientists involved, the committee gradually lost its initial impetus and existed only in theory until 1946 when Yerkes took the time to write the appropriate letters to formally terminate the committee. For information on the committee see a newspaper clipping from the *New York Times*, December 28, 1930, and Yerkes to Chapman, March 4, 1931. RMY, Box 72, Folder 1385. For information on the gradual demise of the committee, see RMY, Box 72, Folder 1384. For more on the establishment of the Albert National Park, see Jeannette Eileen Jones, *In Search of Brightest Africa: Reimagining the Dark Continent in American Culture, 1884–1936* (Athens: University of Georgia Press, 2010).

77. RMY, Box 131, Folder 2237.

78. Harold Bingham to Yerkes, September 3, 1929, Harold Bingham to Raymond Dodge, September 9, 1929. RMY, Box 4, Folder 69.

79. Harold Bingham to Raymond Dodge, September 9, 1929. RMY, Box 4, Folder 69.

80. The only exception was a few photographs taken at close range of the back of a male gorilla that had been shot.

81. Harold Bingham to Richard Sparks, February 15, 1932. RDS, Folder 4.

82. Unidentified friend to Ada Yerkes, March 18, 1930, emphasis is in the original. RMY, Box 4, Folder 70.

83. Yerkes to James R. Angell, March 25, 1930. RMY, Box 4, Folder 70.

84. Richard Sparks to Harold Bingham, October 10, 1931. RDS, Folder 4.

85. There are many letters that contain information about the relationship between Bingham and Yerkes. For one example, see Yerkes to James R. Angell, March 25, 1930. RMY, Box 4, Folder 70.

86. For a history of the laboratory, see Dewsbury, *Monkey Farm*.

87. Yerkes, "Foreword," iii.

88. For more on the use of primates in the laboratory, see Elizabeth Hanson, "How Rhesus Monkeys Became Laboratory Animals," in *Centennial History of the Carnegie Institution of Washington* 5 (Department of Embryology), ed. Jane Maienschein, Marie Glitz, and Garland E. Allen (New York: Cambridge University Press, 2005), 63–82.

89. Henry W. Nissen, "A Field Study of the Chimpanzee: Observations of Chimpanzee Behavior and Environment in Western French Guinea," *Comparative Psychology Monographs* 8, no. 1 (1931): 16.

90. Nissen, "A Field Study of the Chimpanzee," 2.

91. Yerkes to Henry W. Nissen, March 14, 1930. RMY, Box 36, Folder 689.

92. Henry W. Nissen to Yerkes, February 20, 1930. RMY, Box 36, Folder 688.

93. Henry W. Nissen to Otto L. Tinklepaugh, March 3, 1930. RMY, Box 36, Folder 688.

3. Control, Repetition, and Objectivity

1. Clarence R. Carpenter, "The Effect of Complete and Incomplete Gonadectomy on the Behavior and Morphological Characters of the Male Pigeon" (PhD disserta-

tion, Stanford University, 1931); and Clarence R. Carpenter, "The Relation of the Male Avian Gonad to Responses Pertinent to Reproductive Phenomena," *Psychological Bulletin* 29, no. 7 (1932): 509–27.

2. Clarence Ray Carpenter, "Psychobiological Studies of Social Behavior in Aves: The Effect of Complete and Incomplete Gonadectomy on the Primary Sexual Activity of the Male Pigeon" *Journal of Comparative Psychology* 16 (1933): 30.

3. Ibid, 36.

4. For a detailed examination of this committee, including Robert Yerkes's influence on the research it funded, see Adele E. Clarke, *Disciplining Reproduction: Modernity, American Life Sciences, and "The Problems of Sex"* (Berkeley: University of California Press, 1998). See also Wade Pickren, "Robert Yerkes, Calvin Stone, and the Beginning of Programmatic Sex Research, 1921–1930," *American Journal of Psychology* 110, no. 4 (1997): 603–19.

5. For more on the history of the IRTA, see Joel B. Hagen, "Problems in the Institutionalization of Tropical Biology: The Case of the Barro Colorado Island Biological Island," *History and Philosophy of the Life Sciences* 12 (1990): 225–47, esp. 227–31.

6. See Megan Raby, "Making Biology Tropical: American Science in the Caribbean, 1898–1963" (PhD dissertation, University of Wisconsin Madison, 2013), chapter 4.

7. F. Chapman, *My Tropical Air Castle: Nature Studies in Panama* (New York, D. Appleton-Century, 1929), 1. For more on Chapman's role in the history of ornithology, see E. Mayr, "Epilogue: Materials for a History of American Ornithology," in *Ornithology: From Aristotle to the Present,* ed. E. Stresemann (Cambridge, MA: Harvard University Press, 1975), 369–73.

8. Chapman, *My Tropical Air Castle,* 12. Chapman also stated that the climate was favorable and the island "healthful" due to the lack of malaria. Hagen has also noted that Chapman's works served to promote Barro Colorado as a site for biological research. See Hagen, "Problems in the Institutionalization of Tropical America," 233.

9. Donna Haraway has noted the role of Chapman's letter to Yerkes in his welcome of Carpenter's application for a postdoctoral position at Yale Laboratories of Primate Biology. See Donna Haraway, *Primate Visions: Gender, Race and Nature in the World of Modern Science* (New York: Routledge, 1989), 87–88.

10. Clarence R. Carpenter, "A Field Study of the Behavior and Social Relations of Howling Monkeys (*Aloatta palliata*)," *Comparative Psychology Monographs* 102, no. 2 (1934), 17

11. Ibid., 20–21.

12. Haraway, *Primate Visions,* 88; and E. Mayr, "Epilogue," 365–96.

13. Carpenter, "A Field Study of the Behavioral and Social Relations," 62.

14. Ibid., 101–2.

15. See, for example, Francis H. Herrick, "An Eagle Observatory," *Auk* XLI (1924) and Guy A. Bailey, "The Trailer-Blind de Luxe," *Bird Lore,* vol. 24, no. 6 (1922).

16. Carpenter, "A Field Study of the Behavior and Social Relations," 25.

17. Ibid., 26–27.

18. Ibid., 33.

19. Ibid., 20. Here Carpenter relegates the function of defense to "no useful function

whatever" in an attempt to point out the limited role vocalizations were understood to play in animal societies. Poor expression such as this was a cause of friction between Carpenter and Yerkes, who refused to let Carpenter make observations at Yale Laboratories of Primate Biology until Carpenter had spent much time reluctantly revising his 1934 monograph. See Yerkes to Henry Nissen, March 5, 1934. RMY, Box 36, Folder 693.

20. Carpenter, "A Field Study of the Behavior and Social Relations," 110.

21. Ibid., p, 111.

22. Wolfgang Köhler, *The Mentality of Apes* (New York: Harcourt, 1925).

23. Robert M. Yerkes and Ada Yerkes, *The Great Apes* (New Haven: Yale University Press, 1929), 308. Emphasis is in original.

24. For information about Wolfgang Köhler, see Marion Thomas, "Rethinking the History of Ethology: French Animal Behavior Studies in the Third Republic (1870–1950)" (PhD dissertation, Manchester University, 2003).

25. In addition to *The Great Apes,* see Robert M. Yerkes, *Almost Human* (New York: The Century Co., 1925), and Robert M. Yerkes and Blanche Learned, *Chimpanzee Intelligence and Its Vocal Expressions* (Baltimore: Williams and Wilkins, 1925).

26. Clarence R. Carpenter, "Approaches to Studies of the Naturalistic Communicative Behavior in Nonhuman Primates," in *Approaches to Animal Communication,* ed. Thomas A. Sebeok and Alexandra Ramsey (The Hague, the Netherlands: Mouton, 1969), 45–46.

27. Wallace Craig, "The Voices of Pigeons Regarded as a Means of Social Control," *American Journal of Sociology* 14 (1908): 86–100. For secondary sources on Craig, see Richard W. Burkhardt Jr., "Charles Otis Whitman, Wallace Craig and the Biological Study of Animal Behavior in the United States, 1898–1925," in *The American Development of Biology,* ed. R. Rainger, Keith Benson, and Jane Maienschein (New Brunswick, NJ: Rutgers University Press, 1988), 185–218; and Richard W. Burkhardt Jr., "Ethology, Natural History, the Life Sciences and the Problem of Place," *Journal of the History of Biology* 32 (1999): 489–508; Gregg Mitman and Richard W. Burkhardt Jr., "Struggling for Identity: The Study of Animal Behavior in America, 1930–1945," in *Expansion of American Biology,* ed. R. Rainger, Jane Maienschein, and Keith Benson (New Brunswick, NJ: Rutgers University Press, 1991), 164–94; and Richard W. Burkhardt Jr., *Patterns of Behavior: Konrad Lorenz, Niko Tinbergen and the Founding of Ethology* (Chicago: University of Chicago Press, 2005). Haraway has also noted that Craig influenced Carpenter; see Donna Haraway, "Signs of Dominance: From a Physiology to a Cybernetics of Primate Society, C.R. Carpenter, 1930–1970," *Studies in History of Biology* 6 (1983): 141.

28. Craig, "The Voices of Pigeons," 87.

29. It is interesting to note that, in 1969, Carpenter pointed out that his doctoral research confirmed Craig's results. However, Carpenter did not cite Craig's paper in either his dissertation or the immediate publications that resulted from it. This can probably be explained by the fact that Craig's work was becoming increasingly, although still narrowly, cited during the 1930s. Thus, Carpenter came to realize the connections between his own work and that of Craig's shortly after completing his

dissertation and went on to cite Craig's 1908 paper in both his howler and gibbon monographs. I thank Richard W. Burkhardt Jr. for suggesting this explanation for the delay in Carpenter's citation of Craig's work.

30. See Burkhardt Jr., *Patterns of Behavior*, chapter 2, esp. 92–98; and Haraway, *Primate Visions*, 391.

31. Haraway, "Signs of Dominance," 190. In 1942, Carpenter compared the use of vocalizations by primates to decrease aggression to the League of Nations. See Clarence R. Carpenter, "Societies of Monkeys and Apes," *Biological Symposia* 9 (1942): 204.

32. For biographical information, see E. M. Kersey, *Women Philosophers: A Biocritical Sourcebook* (New York: Greenwood Press, 1989), 90–91.

33. See Carpenter, "Approaches to Studies," 46.

34. Craig, "The Voices of Pigeons," 87.

35. Richard W. Burkhardt Jr. has also noted Craig's emphasis on the importance of looking beyond the individual. Burkhardt Jr., *Patterns of Behavior*, 39.

36. Grace de Laguna, *Speech: Its Function and Development* (New Haven: Yale University Press, 1927), 28n2.

37. Ibid., 28.

38. *Speech* was included in the bibliographies of all Carpenter's major works involving primate communication. He also discussed Grace de Laguna's influence in a letter to her daughter; Carpenter to Fredericka de Laguna, January 18, 1969. CRC, Box 8, Folder 10.

39. Harold J. Coolidge to Robert Mearns Yerkes, June 6 1936. RMY Box 12, Folder 199.

40. Carpenter's APE field notebook, 301. CRC Box 5, Folder 7.

41. "Scientists to Live with Jungle Apes," *New York Times*, December 16, 1936.

42. Londa Schriebinger, "Gendered Ape," in *A Question of Identity: Women, Science and Literature*, ed. Marina Benjamin (New Brunswick, NJ: Rutgers University Press, 1993).

43. Harold J. Coolidge to Robert Mearns Yerkes, June 6, 1936. RMY Box 12, Folder 199.

44. "A.P.E: Second Informal Report on the Asiatic Primate Expedition," October 15, 1937.RMY Box 10, Folder 166.

45. Carpenter did not approach primate field studies from the standpoint of one area of specialty; rather, he was eclectic, studying a range of behaviors to get a panoramic view of primates' lives and behaviors. Harold Coolidge to Yerkes, September 24, 1940, and September 27, 1940. RMY, Box 12, Folder 200.

46. Harold Coolidge to Yerkes, September 27, 1940. RMY, Box 12, Folder 200.

47. For works on gender and the field, see Bruce Hevly, "The Heroic Science of Glacier Motion," *Osiris*, vol. 11 (1996): 66–86; and Donald Optiz, "'A Triumph of Brains over Brute:' Women and Science at the Horticultural College, Swanley, 1890–1910," *ISIS* 104 (2013): 30–62.

48. Clarence R. Carpenter, "A Field Study in Siam of the Behavior and Social Relations of the Gibbon (*Hylobates lar*)," *Comparative Psychology Monograph* (1940): 2

49. Carpenter discusses his feeling of hopelessness having failed to observe gibbons after two months in the field in his field notes, 16. CRC, Box 5, Folder 7.

50. Carpenter, "A Field Study in Siam," 20.

51. Ibid., 34.

52. Ibid., 169.

53. For a late nineteenth-century example, see Gregory Radick, *Simian Tongue: The Long Debate about Animal Language* (Chicago: Chicago University Press, 2008), and Jeremy Rich, *Missing Links: The African and American Worlds of R. L. Garner, Primate Collector* (Athens: University of Georgia Press, 2012). The Huxley-Koch study is discussed in Julian Huxley and L. Koch, *Animal Language* (London: Country Life, 1938).

54. I thank Peter Copeland, technical manager at the British Library National Sound Archive, for providing this weight estimate.

55. Carpenter's APE notebook, 220. CRC, Box 5, Folder 7.

56. Ibid., 154.

57. RMY, Box 10, Folder 166; Clarence R. Carpenter, "Behavior and Social Relations of Free-Ranging Primates," *Scientific Monthly* 48 (1939): 325; and Clarence R. Carpenter, "A Field Study in Siam," 180.

58. Clarence R. Carpenter, "Social Behavior of Non-Human Primates," *Physiologie des Societes Animales* 34, no. 3 (1950): 228.

59. Carpenter, "Approaches to Studies," 55.

60. Ibid.

61. Ibid.

62. Carpenter, "A Field Study in Siam," 170.

63. Carpenter, "Field Studies of Free Ranging Primates with Special Emphasis on the Gibbon," page 4. CRC Box 2, Folder 67.

64. Examples of praise for Carpenter as a trailblazer in primatology include Robert Ardrey, *African Genesis: A Personal Investigation into the Animal Origins and Nature of Man* (New York: Atheneum, 1961), 35–36; John P. Scott, *Animal Behavior* (Chicago: University of Chicago Press, 1958); Earnest Hooton, *Man's Poor Relations* (New York: Doubleday, 1942); Stuart Altmann, *Social Communication Among Primates* (Chicago: University of Chicago Press, 1967), 376; Harold J. Coolidge to Carpenter, September 21, 1970, in CRC, Box 7, Folder 38; Irven DeVore's edited volume, *Primate Behavior*, one of two books published in 1965 to demonstrate the galvanization of primate field studies, was also dedicated to Carpenter for his work in the field. I. DeVore, ed., *Primate Behavior: Field Studies of Monkeys and Apes* (New York: Holt, Reinhart and Winston, 1965).

65. See S. Eimerl and I. DeVore, *Primates* (New York, Time-Life Books, 1965); and Nancy Robinson, *Jungle Laboratory: The Story of Ray Carpenter and the Howling Monkeys* (New York: Hastings House, 1973).

4. Capturing Natural Behavior

1. Francis Sumner used the term *second class*; see Robert Kohler, *Labscapes and Landscapes: Exploring the Lab-Field Border in Biology* (Chicago: University of Chicago Press, 2002), 200.

2. Carpenter used the term *retarded status*; see C. R. Carpenter, "Social Behavior of Non-Human Primates," 228.

3. Kohler, *Labscapes and Landscapes*, 213.

4. Ibid., 256.

5. Robert M. Yerkes and Ada Yerkes, "Social Behavior in Infrahuman Primates," in *A Handbook of Social Psychology*, ed. C. Murchison (Worcester, MA: Clark University Press, 1935), 987; Robert M. Yerkes and Ada Yerkes, "Social Behavior in Infrahuman Primates," 982; Robert Ardrey wrote similar critiques of Zuckerman's Monkey Hill study. See Robert Ardrey, *African Genesis: A Personal Investigation into the Animal Origins and Nature of Man* (New York: Atheneum, 1961), 34–35. Yerkes dislike for Zuckerman's research methods also comes through when reading papers in his archive. For example, after briefly courting Zuckerman for a position at Yale Laboratories of Primate Biology in 1934, Yerkes concluded: "He is essentially an enthusiastic, impatient skimmer of the cream of his scientific environment. I think very well of his scientific output, but the two books which he has published show haste, carelessness, over-eagerness to make shrewd guesses, and seeming unwillingness to do the exacting, pains-taking work which is essential to sound scholarship and to lasting valuable contribution." Thus, he opted not to hire Zuckerman. Yerkes to Warren Weaver (Rockefeller Foundation), June 1, 1934. RMY, Box 54, Folder 1046. It is interesting that Nissen had used the term "cream skimming" in reference to Zuckerman one year earlier when contrasting Zuckerman's impatient nature with Carpenter's painstaking work on Barro Colorado. See Henry Nissen to Yerkes, February 13, 1933. RMY, Box 54, Folder 1045. For further evidence, see Yerkes to Alan Gregg, June 27, 1932. RMY, Box 54, Folder 1044. For more on Zuckerman, see Donna Haraway, *Simians, Cyborgs, and Women: The Reinvention of Nature* (London: Free Association Press, 1991), chapter 2.

6. Clarence R. Carpenter, "A Field Study in Siam of the Behavioral and Social Relations of the Gibbon (*Hylobates lar*)," *Comparative Psychology Monograph* (1940): 130–31, 134–35, and 183.

7. See, for example, Clarence R. Carpenter, *Naturalistic Behavior of Nonhuman Primates* (University Park: Pennsylvania State University Press, 1964), 428; and Richard G. Rawlins and Matt J. Kessler, *The Cayo Santiago Macaques: History, Behavior and Biology* (Albany: State University of New York Press, 1986), 14–21.

8. In addition to Carpenter's own publications concerning the breeding of primates, he collected many newspaper clippings about this issue. Examples of such a news article are "Export of Monkeys," *Sunday Standard*, September 26, 1937, and "Monkeys Arrive Here for Breeding Purposes," *Washington Post*, November 14, year not recorded but content of the article would indicate it to have been 1937. CRC, Box 5, Folder 8.

9. Clarence R. Carpenter, "Rhesus Monkeys (*Macaca mulatta*) for American Laboratories," *Science* 92, 2387 (1940): 285. For a secondary source dealing with the use of animals for experimentation, see A. Guerrini, *Experimenting with Humans and Animals: From Galen to Animal Rights* (Baltimore: Johns Hopkins Press, 2003), chapter 6. Donald A. Dewsbury provides an account of the establishment of the American Psychological Association Committee on Precautions in Animal Experimentation in 1925 in response to the antivivisection movement. Yerkes was the chair of this committee, if somewhat reluctantly. Donald A. Dewsbury, "Early Interactions between Animal Psychologists and Animal Activists and the Founding of the APA Committee

on Precautions in Animal Experimentation," *American Psychologist* 45, no. 3 (1990): 315–27.

10. Carpenter, "Rhesus Monkeys (Macaca mulatta) for American Laboratories," 284.

11. CRC, Box 5, Folder 8.

12. Clarence R. Carpenter, "Breeding Colonies of Macaques and Gibbons on Santiago Island, Puerto Rico" in *Breeding Primates, Proceedings of the International Symposium on Breeding Non-Human Primates for Laboratory Use, Berne, 28–30 June 1971*, ed. W. I. B. Beverdige (Basel, Switzerland: S. Karger, 1972), 77–78.

13. This focus on recognition of individuals reflected the beginning of an emerging trend in field primatology. See A. F. Richard and S. R. Schulam, "Sociobiology: Primate Field Studies," *Annual Review of Anthropology* 11 (1982): 231–55, esp. 236 and 244.

14. CRC, Box 5, Folder 8; and Donna Haraway, *Primate Visions: Gender, Race and Nature in the World of Modern Science* (New York: Routledge, 1989), 85–86.

15. Clarence R. Carpenter, "Sexual Behavior of Free Ranging Rhesus Monkeys (*Macaca mulatta*): I. Specimens, Procedures and Behavioral Characteristics of Estrus," *The Journal of Comparative Psychology* 33 (1942): 141.

16. Clarence R. Carpenter, "Sexual Behavior of Free Ranging Rhesus Monkeys (*Macaca mulatta*): II. Periodicity of Estrus, Homosexual, Autoerotic and Non-Conformist Behavior," *Journal of Comparative Psychology* 33 (1942): 150.

17. Ibid., 152.

18. Jan 21, 1939. RMY, Box 10 Folder 166. Emphasis is in the original.

19. For examples of literature that illustrates the shy demeanor of the gorilla, see Carl Akeley, *In Brightest Africa* (New York: Garden City, 1923); and M. H. Bradley, *On the Gorilla Trail* (New York: D. Appleton, 1922).

20. It was this reputation that later led reviewers of George Schaller and John Emlen's 1959 study, such as Carpenter himself, to state that such a project did not have a great chance of success. In fact, Carpenter rated their proposal only a three, or acceptable, due to fears that the cost, training, and manpower required for such a study had all been underestimated. Schaller was also proposing to take his wife on the expedition, to which Carpenter commented, "I personally distrust the man-wife combination." Despite Carpenter's fears, Schaller and Emlen, and their wives, undertook the New York Zoological Society African Primate Expedition from 1959–61. This lengthy field project was very successful, and publication of their study of gorilla behavior was met with widespread praise, including public and private admiration from Carpenter. For information on Schaller and Emlen's gorilla field study, see G. Schaller, *The Mountain Gorilla: Ecology and Behavior* (Chicago: University of Chicago Press, 1963); J. T. Emlen and G. Schaller, "In the Home of the Mountain Gorilla," *Animal Kingdom* 63, no. 3 (1960): 98–108; and J. T. Emlen and G. Schaller, "Distribution and Status of the Mountain Gorilla," *Zoologica, Scientific Contributions of the New York Zoological Society* 45, no. 1 (1960): 41–52. Carpenter's view of the project as proposed by Schaller and Emlen, including his statement about husband and wives in the field, can be found in CRC, Box 12, Folder 3.

21. Yerkes to Adolph Schultz, January 13, 1950. RMY, Box 43, Folder 829.

22. See, for example, Robert M. Yerkes, "Gorilla Census and Study," *Journal of Mammalogy* 32, no. 4 (1951): 429–36.

23. Notes from the "Gorilla Trail" in RMY, Box 65, Folder 1226.

24. Yerkes, "Gorilla Census and Study," 434.

25. For an overview of Benchley's work at the zoo in the form of a children's book, see Margaret Poynter, *The Zoo Lady: Belle Benchley and the San Diego Zoo* (Minneapolis: Dillon Press, 1980); and Emily Hahn, *Eve and the Apes* (New York: Weidenfeld & Nicolson, 1988), 9–24.

26. Margaret Rossiter, *Women Scientists in America: Before Affirmative Action, 1940–1972* (Baltimore: Johns Hopkins University Press, 1995), 247; see also Margaret Rossiter, *Women Scientists in America: Struggles and Strategies to 1940* (Baltimore: Johns Hopkins University Press, 1982), 265.

27. Belle Jennings Benchley, *My Friends, The Apes* (Boston: Little, Brown, 1942).

28. Belle Jennings Benchley, *My Animal Babies* (London: Faber and Faber, 1946).

29. Benchley to Ada Yerkes, March 18, 1958; Benchley to Ada Yerkes, February 13, 1957; Benchley to Ada Yerkes, March 18, 1958; Benchley to Ada Yerkes, August 20, 1958. RMY, Box 4, Folder 62.

30. Poynter, *The Zoo Lady*, 30–36.

31. Harold Bingham to Richard Sparks, September 1, 1931. RDS, Folder 4.

32. Newspaper clipping featuring the dinner act from March 15, 1952 (the newspaper's title is unfortunately not recorded). RMY, Box 66, Folder 1237.

33. See "Steps in Gorilla Training Program." RMY, Box 65, Folder 1226. A series of correspondences concerning Kelly's failure to conduct the study in a timely manner and ultimate failure to publish the study were exchanged between Benchley and Ada Yerkes. See, for example, Belle Benchley to Ada Yerkes, August 8, 1958. RMY, Box 4, Folder 62.

34. Yerkes to Carpenter February 10, 1934, and Carpenter to Yerkes April 7, 1934. RMY, Box 10, Folder 163.

35. Clarence R. Carpenter, "An Observational Study of Two Captive Mountain Gorillas (*Gorilla beringei*)," *Human Biology* 9, no. 2 (1937): 175–96.

36. Carpenter, "An Observational Study," 193.

37. Yerkes to Carpenter May 2, 1936. RMY, Box 10, Folder 165.

38. "Memorandum on Prospective Conference with Mr. Yakima Canutt," March 31, 1954. RMY, Box 66, Folder 1235.

39. Yerkes to Rubin (MGM), March 4, 1953. RMY, Box 66, Folder 1235.

40. Gregg Mitman has discussed the use of gorillas in films, including the proposed use of a fenced enclosure within which it was expected that the gorillas would continue to behavior as they would in a "wild state." See Gregg Mitman, *Reel Nature: America's Romance with Wildlife on Film* (Cambridge, MA: Harvard University Press, 1999), 55–58.

41. Yerkes to Coolidge, March 21, 1955. RMY, Box 12, Folder 200.

42. Belle Benchley to Yerkes, October 1, 1953. RMY, Box 66, Folder 1235.

43. Both quotes are in Belle Benchley's report to Yerkes. RMY, Box 66, Folder 1235.

44. Carpenter's field techniques directly affected the work of John P. Scott. See

John P. Scott, "Social Behavior, Organization and Leadership in a Small Flock of Domestic Sheep," *Comparative Psychology Monographs* 18, no. 4 (1945): 1–29; and Gregg Mitman, "Dominance, Leadership and Aggression: Animal Behavior Studies during the Second World War," *Journal of the History of the Behavioral Sciences* 26 (1990): 10.

45. Furthermore, the vast majority of committee members also shared Carpenter's concerns about field methods. A subcommittee was devoted to the development of instrumentation for the field and comprised of Donald Griffin from Cornell University, John B. Calhoun from Johns Hopkins, and John L. Fuller from Jackson Hole Memorial Laboratories. See Committee for the Study of Animal Societies under Natural Conditions, 1948, 1949, and 1951 Reports. RMY, Box 70, Folder 1337.

46. Schneirla argued that the observer should use practices such as repetition, effective note taking, and sound and motion recording technology to ensure what he called "observational control." T. C. Schneirla, "The Relationship between Observation and Experimentation in the Field Study of Behavior," *Annals of the New York Academy of Sciences* 51, no. 6 (1950): 1022–44.

47. This interpretation of the observer as active is present throughout the paper, but see Schniela, "The Relationship between Observation and Experimentation," 1025, for this specific example.

48. Ibid., 1025, 1042.

49. Ibid., 1023, 1025.

50. Carpenter, "Social Behavior of Non-Human Primates," 228.

51. J. T. Emlen and F. W. Lorenz, "Pairing Responses of Free-Ranging Valley Quail to Sex Hormone Pellet Implants," *Auk*, 59 (1942): 369–78.

52. J. T. Emlen, "Techniques for Observing Bird behavior under Natural Conditions," *Annals for the New York Academy of Sciences*, 51 no. 6 (1950): 1103.

53. Niko Tinbergen, "Social Releasers and the Experimental Method Required for Their Study," *Wilson Bulletin* 60, no. 1 (1948): 6–51; See also Richard W. Burkhardt Jr., *Patterns of Behavior*, especially chapters 4 and 6.

54. Gregg Mitman, *The State of Nature: Ecology, Community and American Social Thought, 1900–1950* (Chicago: University of Chicago Press, 1992), 4.

55. E. M. Banks, "Warder Clyde Allee and the Chicago School of Animal Behavior," *Journal of the History of the Behavioral Sciences* 21 (1985): 348–49.

56. Edwin P. Willems and Harold L. Raush, "Introduction," in *Naturalistic Viewpoints in Psychological Research*, ed. Edwin P. Willems and Harold L. Raush (New York: Holt, Rinehart and Winston, 1969), 8.

57. William Mason, "Back to Nature," *Contemporary Psychology* 15 (1970): 89.

58. Mason, "Back to Nature," 89. Emphasis is in the original.

59. Edwin P. Willems, "Planning a Rationale for Naturalistic Research," in *Naturalistic Viewpoints*, 49–50.

60. Ibid., 46 and 66. See also Edwin P. Willems and Harold L. Rausch, "Interpretations and Impressions," in *Naturalistic Viewpoints*, 275. Within the history of ethology, one could identify "instinct" as a similarly troublesome word. Various theories and critiques of instinct are discussed in Richard W. Burkhardt Jr., *Patterns of Behavior: Konrad Lorenz, Niko Tinbergen and the Founding of Ethology* (Chicago: University of Chicago Press, 2005).

61. E. W. Menzel, "Naturalistic and Experimental Approaches to Animal Behavior," in *Naturalistic Viewpoints*, 80.

62. Ibid., 117.

63. Ibid., 118.

64. The use of naturalness to form an identity for scientific research can also be seen in the history of ethology. Burkhardt has discussed how the study of natural behavior was central to the identity of ethologists; see Burkhardt, *Patterns of Behavior*, 10.

5. Modern Primatology

1. Sparks to Bingham, October 10, 1931, in RDS, Folder 4.

2. Clarence R. Carpenter's review of John Emlen's proposal to study the "Behavior and Ecology of the Mountain Gorilla in its Native Habitat." CRC Box 12, Folder 3.

3. George Schaller, "The Mountain Gorilla," *The New Scientist* no. 268 (1962):16.

4. John O'Reilly, "The Amiable Gorilla," *Sports Illustrated*, June 20, 1960.

5. George B. Schaller, *The Mountain Gorilla: Ecology and Behavior* (Chicago: University of Chicago Press, 1963), 5. It is noteworthy that his acknowledgments also included recognition of his wife, Kay, who served the role of "secretary, field hand, critic, cook, companion—a combination of jobs only willingly taken on by a wife" (6) Thus, in contrast to Carpenter's monographs, Schaller did not disguise the presence of his wife in the field; indeed Schaller even included photos of her observing gorillas.

6. Schaller, *Mountain Gorilla*, 20.

7. Ibid., 18.

8. Ibid., 19.

9. For more on Fossey's life and research, see Georgina M. Montgomery, "Fossey, Dian," *The New Dictionary of Scientific Biography* (Florence, KY: Gale Cengage, 2007).

10. For a history of the concept of infanticide, see Amanda Rees, *The Infanticide Controversy: Primatology and the Art of Field Science* (Chicago: University of Chicago Press, 2009).

11. Dian Fossey, "Making Friends with Mountain Gorillas," *National Geographic*, January 1970; and Dian Fossey, "More Years with Mountain Gorillas," *National Geographic* January 1971.

12. Dian Fossey, "The Behaviour of the Mountain Gorilla" (PhD dissertation, University of Cambridge, 1976); and Dian Fossey, *Gorillas in the Mist* (Boston: Houghton Mifflin, 1983).

13. Farley Mowat, *Woman in the Mists: The Study of Dian Fossey and the Mountain Gorillas of Africa* (New York: Warner Books, 1987), 59.

14. Fossey, *Gorillas in the Mist*, 206.

15. Dian Fossey is quoted in Susan Chira, "Symposium Sees Ways Apes Resemble Humans" *New York Times*, May 17, 1982, B2.

16. Fossey, *Gorillas in the Mist*, 34–35.

17. Mowat, *Woman in the Mists*, 165.

18. Many media outlets claimed "poachers" to be suspects in Fossey's death. See, for example, "Zoologist Is Slain in Central Africa," *New York Times*, December 29, 1985, 15.

19. Donna Haraway, *Primate Visions: Gender, Race and Nature in the World of Modern Science* (New York: Routledge, 1989), 265.

20. Gregg Mitman, "Pachyderm Personalities: The Media of Science, Politics, and Conservation," in *Thinking with Animals: New Perspectives on Anthropomorphism*, ed. Lorraine Daston and Greg Mitman (New York: Columbia University Press, 2005), 190.

21. Montgomery, "Fossey, Dian."

22. James Krasner, "'Ape Ladies' and Cultural Politics: Dian Fossey and Biruté Galdikas," in *Natural Eloquence: Women Reinscribe Science*, ed. Barbara T. Gates and Ann B. Shteir (Madison: University of Wisconsin Press, 1997).

23. Konrad Lorenz, *On Aggression* (New York: Harcourt, Brace & World, 1966). For further context, see Nadine Weidman, "Popularizing the Ancestry of Man: Robert Ardrey and the Killer Instinct," *ISIS* 102 (2011): 269–99.

24. Amanda Rees, "Reflections on the Field: Primatology, Popular Science and the Politics of Personhood," *Social Studies of Science* 37 (2007): 881–907.

25. "Animal Behaviorist Finds Chimpanzees Take Others' Lives," *New York Times*, April 20, 1978, A19.

26. Jane Goodall, "Watching, Watching, Watching," *New York Times*, September 15, 1977, 24.

27. Roger Fouts, *Next of Kin: What Chimpanzees Have Taught Me About Who We Are* (New York: A Living Planet Book, 1997); and Herbert Terrace, *Nim, A Chimpanzee Who Learned Sign Language* (New York: Columbia University Press, 1987).

28. Bayard Webster, "Man as Ape Was the Key to Filming 'Greystoke,'" *New York Times*, March 25, 1984, H1.

29. Benedict Carey, "Washoe, A Chimp of Many Words, Dies at 42," *New York Times*, November 1, 2007, A15.

30. Francine Patterson, "Conversations with a Gorilla," *National Geographic*, vol. 154, no. 4 (October 1978): 438–65.

31. See, for example, the front cover of *National Geographic*, vol. 128, no. 6 (December 1965). Donna Haraway analyzes these images and others in "Apes in Eden, Apes in Space: Mothering as a Scientist for National Geographic" in *Primate Visions*. For a broader examine of the journal as a whole see Catherine A. Lutz and Jane L. Collins, *Reading National Geographic* (Chicago: University of Chicago Press, 1993).

32. Interview with Mike Wilson, Minneapolis, 2007.

33. Meredith F. Small, *Female Primates: Studies by Women Primatologists* (New York: A. R. Liss, 1984).

34. For more on the number of women in primatology, see Carol Jahme, *Beauty and the Beasts: Woman, Ape, and Evolution* (New York: Soho Press, 2001). The percentage of women in primatology has and does vary depending on the institution and type of position being examined. See Linda Marie Fedigan, "Science and Successful Female: Why Are There So Many Women Primatologists," *American Anthropologist*, vol. 96, no. 3 (1994): 529–40.

35. Frederick Weisman, *Primate* (Zipporah Films, 1974).

36. Irven DeVore, ed., *Primate Behavior: Field Studies of Monkeys and Apes* (New York, Holt, Rinehart and Winston, 1965); and Allen M. Schrier, Harry F. Harlow, and Fred Stollnitz, *Behavior of Nonhuman Primates: Modern Research Trends* (New York: Academic Press, 1965).

37. DeVore, *Primate Behavior*, viii.
38. Ibid., 9.
39. Ibid., 27.
40. Ibid., 621–22.
41. Stuart Altmann, "Primate Behavior in Review," *Science* 10 (1965).

6. Inclusion and Indigenous Researchers

1. Peter M. Kappeler, Carel P. van Schaik, and David P. Watts, "The Values and Challenges of Long-Term Field Studies," in *Long-Term Field Studies of Primates*, ed. Peter M. Kappeler and David P. Watts (Berlin: Springer, 2012), 3–18.

2. For a discussion of these issues regarding Dian Fossey and Biruté Galdikas, see J. Krasner, "'Ape Ladies' and Cultural Politics: Dian Fossey and Biruté Galdikas," in *Natural Eloquence: Women Reinscribe Science*, ed. Barbara T. Gates and Ann B. Shteir (Madison: University of Wisconsin Press, 1997), 237–51.

3. Stuart A. Altmann and Jeanne Altmann, *Baboon Ecology: African Field Research* (Chicago: University of Chicago Press, 1970), 17–18.

4. In addition to *Baboon Ecology*, see Stuart A. Altmann and S. S. Wagner, "Estimating Rates of Behavior from Hansen Frequencies," *Primates* 11 (1970): 181–83.

5. Altmann and Altmann, *Baboon Ecology*, 81–83.

6. Moringe S. Ole Parkipuny and Dhyani J. Berger, "Maasai Rangelands: Links between Social Justice and Wildlife Conservation," in *Voices from Africa: Local Perspectives on Conservation*, ed. David Lewis and Nick Carter (Washington, DC: WWF, 1993), 113–14.

7. For more on the debates concerning how Amboseli National Park should be run, especially under ever-increasing tourism pressure, see David Western and R. Michael Wright, eds., and Shirley Strum, associate ed., *Natural Connections: Perspectives in Community-Based Conservation* (Washington, DC: Island Press, 1994); David Western, *In the Dust of Kilimanjaro* (Washington, DC: Island Press, 1997); and Richard Leakey and Virginia Morell, *Wildlife Wars: My Fight to Save Africa's Natural Treasures* (New York: St. Martin's Press, 2001). There have also been numerous articles in the *New York Times* on this topic. See Jane Perlez, "Only Radical Steps Can Save Wildlife in Kenya, Leakey Says," *New York Times* May 23, 1989; Jane Perlez, "An African Park in Peril," *New York Times*, May 19, 1991; and James C. McKinley Jr., "Kimana Tikondo Group Ranch Journal; Warily, the Masai Embrace the Animal Kingdom," *New York Times*, March 13, 1996.

8. For a discussion of the problems surrounding the use and understandings of the term *Maasai*, see Dorothy L. Hodgson, *Once Intrepid Warriors: Gender, Ethnicity, and the Cultural Politics of Maasai Development* (Bloomington: Indiana University Press, 2001); Peter Rigby, *Cattle, Capitalism, and Class: Ilparakuyu Maasai Transformations* (Philadelphia: Temple University Press, 1992); Dorothy L. Hodgson, ed., *Rethinking Pastoralism in Africa: Gender, Culture and the Myth of the Patriarchal Pastoralist* (Athens: Ohio University Press, 2000); Thomas Spear and Richard Waller, eds., *Being Maasai: Ethnicity and Identity in East Africa* (Athens: Ohio University Press, 1993); Joan N. Knowles and D. P. Collett, "Nature as Myth, Symbol and Action: Notes

towards a Historical Understanding of Development and Conservation in Kenyan Maasailand," *Africa* 59, no. 4 (1993): 434–60. For work examining the nineteenth century, see Christian Jennings, "Beyond Eponymy: The Evidence for Loikop as an Ethnonym in Nineteenth-Century East Africa," *History in Africa* 32 (2005): 199–220, and Christian Jennings, "Scatterlings of East Africa: Revisions of Parakuyo Identity and History, c. 1830–1926" (PhD dissertation, University of Texas-Austin, 2005).

9. Charis Thompson, "When Elephants Stand for Competing Philosophies of Nature: Amboseli National Park, Kenya," in *Complexitites: Social Studies of Knowledge Practices*, ed. John Law and Annemarie Mol (Durham: Duke University Press, 2002). See also W. K. Lindsay, "Integrating Parks and Pastoralists: Some Lessons from Amboseli" in *Conservation in Africa: People, Policies and Practice*, ed. David Anderson and Richard H. Grove (Cambridge: Cambridge University Press, 1987).

10. For more on the challenges of gaining funding for long-term projects, see Kappeler, van Schaik, and Watts, "The Values and Challenges," 3–18; and Susan C. Alberts and Jeanne Altmann, "The Ambolsei Baboon Research Project: 40 Years of Continuity and Change," in *Long-Term Field Studies of Primates*, ed., Peter M. Kappeler and David P. Watts (Berlin: Springer, 2012), 261–87.

11. Stuart A. Altmann, "Field Observations on a Howler Monkey Society," *Journal of Mammology* 40, no. 3 (1959): 317–30.

12. Stuart Altmann, "Walter Reed and After," unpublished letter sent from Stuart Altmann to Georgina M. Montgomery, June 2007; Stuart A. Altmann, "Sociobiology of Rhesus Monkeys. II: Stochastics of Social Communication" *Journal of Theoretical Biology* 8 (1962): 490–522; Stuart A. Altmann, "A Field Study of the Sociobiology of Rhesus Monkeys, *Macaca mulatta*," *Annals of the New York Academy of Sciences* 102 (1962): 338–435.

13. For more on Jeanne Altmann's career, see Jeanne Altmann, "Motherhood, Methods and Monkeys: An Intertwined Professional and Personal Life," in *Leaders in Animal Behavior: The Second Generation*, ed. Lee Drickamer and Donald Dewsbury (Cambridge: Cambridge University Press, 2009), 39–57.

14. Despite this outcome, the 1974 paper did not spring from feminist roots. Written in 1970, Jeanne Altmann had yet to develop the feminist perspective that would characterize aspects of her later professional persona. Nevertheless, the fact that the paper dealt such a powerful blow to bias in animal behavior observation has led to Jeanne Altmann becoming a regular case study in historical and philosophical scholarship concerning feminism and science. See for example, Donna Haraway, *Primate Visions: Gender, Race, and Nature in the World of Modern Science* (New York: Routledge, 1989); Elisabeth A. Lloyd, "Science and Anti-science: Objectivity and Its Real Enemies," in *Science, Politics and Evolution*, ed. Elisabeth A. Lloyd (Cambridge: Cambridge University Press, 2008); see also Linda Marie Fedigan, "The Paradox of Feminist Primatology: The Goddess's Discipline?" in Angela N. H. Creager, Elizabeth Lunbeck, and Londa Schiebinger, *Feminism in Twentieth-Century Science, Technology and Medicine* (Chicago: University of Chicago Press, 2001). My own account of the 1974 paper not stemming directly from feminist roots is based on an interview with Jeanne Altmann conducted on the phone on May 18, 2011.

15. Alberts's publications include Susan C. Alberts, Jason C. Buchan and Jeanne Altmann, "Sexual Selection in Wild Baboons: From Mating Opportunities to Paternity Success," *Animal Behaviour* 72 (2006): 1177–96; Susan C. Alberts, J. A. Hollister-Smith, R. Mututua, S. Sayialel, P. Muruthi, J. K. Warutere, and Jeanne Altmann, "Seasonality and Long-Term Change in a Savannah Environment," in *Seasonality in Primates: Studies of Living and Extinct Human and Non-human Primates*, ed. D. K. Brockman and C. P. van Schaik (Cambridge: Cambridge University Press, 2005); Susan C. Alberts and Jeanne Altmann, "The Evolutionary Past and the Research Future: Environmental Variation and Life History Flexibility in a Primate Lineage," in *Reproductive Fitness in Baboons: Behavioral, Ecological and Life History Perspectives*, ed. L. Swedell and S. Leigh (New York: Kluwer Academic Publishers, 2006), 277–303; Susan C. Alberts and Jeanne Altmann, "Matrix Models for Primate Life History Analysis," in *Primate Life Histories and Socioecology*, ed. P. M. Kappeler and M. E. Pereira (Chicago: University of Chicago Press, 2003); Susan C. Alberts, Heather E. Watts, and Jeanne Altmann, "Queuing and Queue-Jumping: Long-Term Patterns of Reproduction Skew in Male Savannah Baboons, *Papio cynocephalus*," *Animal Behaviour* 65 (2003): 821–40; Susan C. Alberts, Jeanne Altmann, and Michael L. Wilson, "Mate Guarding Constrains Foraging Activity of Male Baboons," *Animal Behaviour* 51 (1996): 1269–77.

16. Interview with Susan Alberts, Amboseli, June 27, 2009. When preparing for conducting interviews in Amboseli, I read methodological works from anthropology including Herbert J. Rubin and Irene S. Rubin, *Qualitative Interviewing: The Art of Hearing Data* (London: Sage Publications, 2005); and Kathleen M. DeWalt and Billie R. DeWalt, *Participant Observation: A Guide for Fieldworkers* (London: Altamira Press, 2002).

17. Like Wagner, Joel E. Cohen, a mathematician at Harvard University, collaborated with Stuart Altmann. See Joel E. Cohen, "Aping Monkeys with Mathematics" in *The Functional and Evolutionary Biology of Primates*, ed. Russell Tuttle (Chicago: Aldine, 1972), 415–36. Stevan J. Arnold and Michael J. Wade were also influenced the Altmanns. See Stevan J. Arnold and Michael J. Wade, "On the Measurement of Natural and Sexual Selection: Theory," *Evolution* 38, no. 4 (1984): 709–19; and Stevan J. Arnold and Michael J. Wade, "On the Measurement of Natural and Sexual Selection: Applications" *Evolution* 38, no. 4 (1984): 720–34. For an example of the Altmann's use of mathematical methods, see Stuart A. Altmann and Jeanne Altmann, "On the Analysis of Rates of Behavior," *Animal Behaviour* 25 (1977): 364–72.

18. Jeanne Altmann's mother-infant research was a particular focus of Donna Haraway's discussion of her work in *Primate Visions*.

19. Examples of testing mathematical models with primate field data include Joel E. Cohen, "Natural Primate Troops and a Stochastic Population Model," *The American Naturalist*, vol. 103, no. 933 (1969): 455–77; and Joel E. Cohen, "Social Grouping and Troop Size in Yellow Baboons," *Proceedings of the Third International Congress of Primatology*, vol. 3 (1970): 58–64.

20. For a discussion of measurement of fitness using long-term field studies, see Stevan J. Arnold, "Laboratory and Field Approaches to the Study of Adaptation," in *Predator-Prey Relationships: Perspectives and Approaches from the Study of Lower*

Vertebrates, ed. Martin E. Feder and George V. Lauder (Chicago: University of Chicago Press, 1986), 169–73.

21. Hans Kummer, review of *Baboon Ecology: African Field Research* by Stuart and Jeanne Altmann in *Science*, vol. 173, no. 4000 (1971): 903.

22. See, for example, Joel Hagen, "The Statistical Frame of Mind in Systematic Biology from "Quantitative Zoology to Biometry," *Journal of the History of Biology* 36, 2 (2003):353–58; and David Sepkoski, "Stephen Jay Gould and Jack Sepkoski, and the Quantitative Revolution in American Paleobiology," *Journal of the History of Biology* 38 (2005): 209–37.

23. Alberts and Altmann, "Matrix Models for Primate Life History Analysis."

24. Altmann and Altmann, *Baboon Ecology*, 17.

25. Fecal samples as a source of DNA is discussed in M. K. Bayes et al., "Testing the Reliability of Microsatellite Typing from Fecal DNA in the Savannah Baboon," *Conservation Genetics* 1 (2000): 173–76.

26. For more on Robert M. Sapolsky, see his *A Primate's Memoir: A Neuroscientist's Unconventional Life among the Baboons* (New York: Simon and Schuster, 2001).

27. Susan Alberts and Greg Wray, "Baboon Susceptibility to a Parasite Reveals Parallels with Humans," *Nature* 460, 7253 (July 16, 2009): 305.

28. Glenn Hausfater, Jeanne Altmann, Stuart Altmann, *Guidebook for the Long-Term Monitoring of Amboseli Baboons and Their Habitat: Definitions, Procedures, and Responsibilities* (published by the authors, 2nd ed., 1982), 27.

29. Interview with Courtney Fitzpatrick, Amboseli, June 26, 2009.

30. Interview with Susan Alberts, Amboseli, June 27, 2009.

31. Interviews with Raphael Mututua, Amboseli, June 23 and 25, 2009.

32. Ibid.

33. Interview with Kinyua Warutere, Amboseli, June 24, 2009.

34. Interviews with Raphael Mututua, Amboseli, June 23 and 25, 2009.

35. *Interview* with Kinyua Warutere, Amboseli, June 24, 2009.

36. Such training of Africans in Western institutions has received criticism from both Western scholars and black African conservationists who recognize Western degrees as colonizing African approaches to animal science and conservation. See M. P. Simbotwe, "African Realities and Western Expectations" in Dale Lewis and Nick Carter, eds., *Voices from Africa* (Washington, DC: World Wildlife Fund, 1993), 18.

37. Jeanne Altmann and Philip Murthuthi, "Differences in Daily Life between Semiprovisioned and Wild-Feeding Baboons" *American Journal of Primatology* 15 (1988): 213–21.

38. J. Altmann, Susan C. Alberts, Susan A. Haines, Jean Dubach, Philip Muruthi, Trevor Coote, Eli Geffen, et al., "Behavior Predicts Genetic Structure in a Wild Primate Group," *Proceedings of the National Academy of Sciences of the United States of America* 93, no. 12 (1996): 5797–801; and Susan C. Alberts, J. A. Hollister-Smith, R. Mututua, S. Sayialel, P. Muruthi, J. K. Warutere, and Jeanne Altmann, "Seasonality and Long-Term Change in a Savannah Environment," in *Seasonality in Primates: Studies of Living and Extinct Human and Non-human Primates*, ed. D. K. Brockman and C. P. van Schaik (Cambridge: Cambridge University Press, 2005).

Conclusion

1. Daniel Bergner, *What Do Women Want? Adventures in the Science of Female Desire* (New York: Harper Collins Press, 2013).

2. Elaine Blair, "I'll Have What She Is Having," in Sunday Book Review, *New York Times,* June 16, 2013.

3. Randy Thornhill and Craig Palmer, *The Natural History of Rape* (Cambridge, MA: MIT Press, 2000) and Cheryl Brown Travis, ed., *Evolution, Gender, and Rape* (Cambridge, MA: MIT Press, 2003).

4. Jeff Hood, *The Silver Back Gorilla Syndrome: Transforming Primitive Man* (Santa Fe, NM: Adventures in Spirit Publications, 1999). For a discussion of how assumptions become so pervasive they go unquestioned, see Roger Lancaster, *The Trouble with Nature: Sex in Science and Popular Culture* (Berkeley: University of California Press, 2003). See also Erika Lorraine Milam, "Making Males Aggressive and Females Coy: Gender across the Animal-Human Boundary," *SIGNS*, vol. 37, no. 4 (2012): 935–59.

5. The announcement was covered by various news outlets including the *New York Times*. James Gorman, "U.S. to Begin Retiring Most Research Chimps," *New York Times,* June 26, 2013.

6. Paola Cavalieri and Peter Singer, eds., *The Great Ape Project: Equality beyond Humanity* (New York: St. Martin's Press, 1994).

7. For more on *Project Nim* and *One Small Step,* see http://www.project-nim.com and http://www.spacechimps.com.

8. "Unconditional Love," *This American Life,* aired on October 24, 2010.

9. See Shirley Strum and Linda Marie Fedigan, eds., *Primate Encounters: Models of Science, Gender, and Society* (Chicago: University of Chicago Press, 2000).

10. Linda Marie Fedigan, "The Paradox of Feminist Primatology: The Goddess's Discipline?" in *Feminism in Twentieth-Century Science, Technology and Medicine,* ed. Angela N. H. Creager, Elizabeth Lunbeck, and Londa Schiebinger (Chicago: University of Chicago Press, 2001), 46–72; and Linda Marie Fedigan, "Science and the Successful Female: Why There Are So Many Women Primatologists," *American Anthropologist* 96, no. 3 (1994): 529–40. For a more conservative estimate of the number of women in primatology, see Elsa Addessi, Marta Borgi, and Elisabetta Palagi, "Is Primatology an Equal-Opportunity Discipline," *PLoS ONE* 7, no. 1 (2012): 1–6.

11. For an example of a scientific field facing very large data sets, see Bruno J. Strasser, "The Experimenter's Museum: GenBank, Natural History, and the Moral Economies of Biomedicine," *ISIS* 102 (2011): 60–69. See also Bernadette Bensaude-Vincent, "A Historical Perspective on Science and Its 'Others,'" *ISIS* 100 (2009): 359–68, esp. 363. I thank Kristoffer Whitney for bringing the Bensaude-Vincent reference to my attention. For more on the PLHD, see Karen B. Strier et al., "The Primate Life History Database: A Unique Shared Ecological Data Resource," *Methods in Ecology and Evolution*, vol. 1 (2010): 199–211.

Index

Abreu, Rosalia, 29, 35–37, 47, 122, 136n33
adventure narratives, 9, 51, 52. *See also* hunter-adventurer
African Primate Field Expedition, 93, 145n20
Akeley, Carl, 41, 47
Akeley, Mary, 47
aggression, 16, 23, 37, 45, 72, 99. *See also* Lorenz, Konrad
Albert National Park, Congo, 47, 81, 93, 139n76. *See also* Akeley, Carl; Akeley, Mary
Alberts, Susan, xi, xii, 1–3, 111–15, 117–18, 120–22, 128. *See also* Amboseli Baboon Project
Allee, Warder Clyde, 86
Allen, Arthur, 67
Altmann, Jeanne, xi, xii, 104, 108–9, 111–15, 117, 121–22, 128; *Baboon Ecology*, 109, 113; *Baboon Mothers and Infants*, 113; and sampling methods, 112, 127, 151n14. *See also* Amboseli Baboon Project
Altmann, Stuart, xi, xii, 93, 104, 108–9, 111, 113–15, 117–18; *Baboon Ecology*, 109, 113; *Foraging for Survival: Yearling Baboons in Africa*, 113. *See also* Amboseli Baboon Project
Amboseli Baboon Project, x, xi, xii, 1, 105–6, 108, 111–17, 119–20, 122, 125–29
Amboseli National Park, Kenya, 107–10, 150n7
American Journal of Primatology, 122
American Museum of Natural History, 47, 55, 84
American Society of Agronomy, 55
American Society of Primatology, 3
anecdotal observations, 21, 23, 26, 103–4. *See also* observation
Animal Kingdom, The (Griffith), 22
animal nature, 18, 125

animal welfare, 32, 41, 74, 125–26
Annals of the New York Academy of Sciences, 84
Anthropoid Experiment Station (Florida), 30, 32
Anthropoid Experiment Station (Tenerife), 59
Asiatic Primate Expedition (A.P.E.), 52–53, 122, 62–63, 65, 68, 70

Baartman, Sara, 16
baboons: graves for, 1–2; and "Monkey Hill," 73; in popular culture, 23; similarity to humans, 1, 129; studies of, xi, 108–9, 111–22. *See also* Alberts, Susan; Altmann, Jeanne; Altmann, Stuart; Amboseli Baboon Project; Fitzpatrick, Courtney; Sapolsky, Robert; Sayialel, Serah; Muruthi, Philip; Mututua, Raphael; Warutere, Kinyua; Zuckerman, Solly
Bachman, George W., 75
Barnard University, 30
Barro Colorado, Panama, 51, 53, 55–59, 61–62, 64–67, 70, 73, 81, 84, 111
Bartholomew, George (Bart), 111
Behaviour, 86, 112
Bekoff, Marc, 125
Benchley, Belle Jennings, 79–81, 83, 90, 120, 122, 126; *My Animal Babies*, 79
Bergner, Daniel, 124
Bingham, Harold, 23–24, 47–52, 78, 81, 92–93, 139n80. *See also* gorilla
birds: collection of, 64; experimental field practices and, 85–86; laboratory studies of, 54–55; pigeons, 53–54, 86, 89; vocalizations of, 59–61. *See also* Howard, Henry Elliot; ornithology; Wallace, Craig
Blair, Elaine, 124

Blumenbach, Johann Friedrich, 12
bonobo: misidentified as chimpanzee, 34; studies of, 20–21, 34
Boston Commercial Bulletin, 16
Bronx Zoo, 84
Brown University, 102
Buddy. *See* Gargantua
Burbridge, Ben, 17, 28, 37–42, 45, 47, 122; *Gorilla: Tracking and Capturing the Ape-Man of Africa*, 38. *See also* Congo (gorilla)
burial: of baboons, 1–2, 120, 125; and Japanese primatology, 8

Carmichael, Leonard, 83
Carnegie Institute, 33, 36, 47–49
Carpenter, Clarence Ray: and Asiatic Primate Expedition, 52, 62–66, 73; and Barro Colorado, 51, 55–58, 73; and captive bird studies, 53–54; and Cayo Santiago, 74–78; interdisciplinary approach, 142n45; and methods and standards for field studies of primates, 6, 53–58, 66–72, 85, 102; studies of communication, 24, 59–62, 67–69; views on naturalness, 73–74, 78, 84–86, 89–90; and zoo studies, 81–82
Carpenter, Mariana, 53, 55, 63, 76, 93
Cartmill, Matt, x
Cayo Santiago, Puerto Rico, 74, 108, 111, 124, 144n8
Central Washington University, 100
Chapman, Frank M., 56, 140n9
Chicago Zoological Society, 122
chimpanzees: and communication, 59–60; as laboratory subjects, 125; in popular culture, 12, 14, 21–22; studies of, 12, 15, 20–21, 28, 30, 32, 34–35, 44–45, 49–51; and tea parties, 25; and tool use, 100. *See also* Abreu, Rosalia; *Joe the Chimpanzee and Other Stories* (anonymous); Goodall, Jane; Nissen, Henry; Tarzan; Yerkes, Robert Mearns
cognitive science, 6
collaboration: across disciplines, xi, 6; between individuals inside and outside traditional scientific community, 4–8, 28, 37; international, 108, 122
Collias, N. E., 84
Columbia University, 49, 75, 100, 125

Committee for Research on Problems of Sex (CRPS), 30, 45, 54–55
Committee for the Study of Animal Societies under Natural Conditions, 84–85, 87, 147n45
communication: playback used in studies of, 67–70; and sign-language studies, 100; studies of, 6, 21, 24, 33, 51, 52, 59–62, 67, 69–70, 103, 111, 140n19, 141n29. *See also* Carpenter, Clarence; De Laguna, Grace; Garner, Richard Lynch; Köhler, Wolfgang; Yerkes, Robert Mearns
Congo (gorilla): as experimental subject, 39, 41; humanlike, 37; and intelligence tests, 40–45; and sexual behavior, 42. *See also* Burbridge, Ben; Sparks, Richard; Yerkes, Robert Mearns
Congorilla: Adventures with the Big Apes and Little People of Africa (Johnson), 18
Coolidge, Harold J., 62–64, 67–68, 83, 93. *See also* Asiatic Primate Expedition
Coolidge, John T., 63. *See also* Asiatic Primate Expedition
copulation, 30, 55, 76, 103, 124
Cornell University, 113
Count Robert of Paris (Scott), 12
Craig, Wallace, 60–62
Cunningham, Alyse, 36

Darwin, Charles, 12–14; *Descent of Man*, 12–14; *The Expression of Emotions in Man and Animals*, 12–14; and gorilla craze, 15–16; *On the Origin of Species*, 12–14
darting, 114–15
De Laguna, Grace, 61–62; *Speech: Its Function and Development*, 61, 142n38. *See also* communication
Deignan, H. G., 63
Delta Regional Primate Center, 87–88
Devore, Irven, 102–3; *Primate Behavior: Field Studies of Monkeys and Apes*, 102–3
Disappointed Traveller, The (Crowquill), 12
Dodge, Raymond, 48
Du Chaillu, Paul Belloni, 16–17, 25, 95; *Explorations and Adventures in Equatorial Africa*, 16–17. *See also* gorilla
Duke University, x, 53, 108, 112, 117

Dupuy, William Atherton, 23
dyadic method, 57, 66

Emlen, J. T., 84–86, 90, 92–95
Engle, Earl T., 75
ethology, 86, 137n49

Fall of Man, The, or The Loves of the Gorilla, A Popular Scientific Lecture upon the Darwinian Theory of Development by Sexual Selection, by a Learned Gorilla (White), 15
Fedigan, Linda Marie, 127
fieldwork: advantages of, 46, 73–74; challenges of, 7, 27–28; experiment used during, 76, 78, 85, 87; long-term, 92, 104, 108, 122, 129; and scientific criteria, 68, 70–71, 85, 92; status of, 51; teamwork and, 53, 64; transnational, 3
film. *See* motion pictures
fitness, 92, 104, 113–14, 152n20. *See also* life history
Fitzpatrick, Courtney, xi, xii, 118
Forest and Stream, 37, 39
Fossey, Dian, 7, 24, 92, 96–98, 100–102, 104, 126, 148n18; and *Gorillas in the Mist*, 97–98. *See also* gorilla
Fouts, Roger, 100
Franklin Farm, 31, 33, 34–35, 49

Galdikas, Birutė, 7
Gardner, Allen, 100
Gardner, Beatrice, 100
Gardner, Richard Lynch, 21–23; and *Speech of Monkeys*, 21
Gargantua (Buddy), 18–20, 24. *See also* gorilla
Genetic Psychology Monographs, 41
gibbons: in popular culture, 22–23, 63; studies of, 24, 52–53, 60, 62–69, 75–76, 85, 142n49. *See also* Asiatic Primate Expedition (A.P.E.); Carpenter, Clarence Ray; Cayo Santiago, Puerto Rico
Goodall, Jane, ix, 7, 92, 98, 100–102, 104, 125–26
gorillas: difficulties studying in wild, 8; and *King Kong*, 22, 24; in popular culture, 15–18, 25, 38–39, 83, 97–98, 125; studies of, 23–25, 37–39, 41–49, 78–84, 91–98, 101, 108;

welfare, 32, 41. *See also* Benchley, Belle Jennings; Bingham, Harold; Congo (gorilla); Cunningham, Alyse; Emlen, John; Fossey, Dian; Gargantua (Buddy); gorilla craze; Schaller, George; Schaller, Kay; Sparks, Richard; Yerkes, Robert Mearns
graves, xvi, 1. *See also* burial
Grey, John Edward, 27
Greystoke: The Legend of Tarzan, Lord of the Apes (film), 100
Griswold, Augustus, Jr., 63
gynecology, 7

habituation, 58, 114. *See also* observation
Hamilton, Iain Douglas, 98
Haraway, Donna, x, 98; *Primate Visions*, x
Harlow, Harry, 102–3, 125; *Behavior of Non-Human Primates: Modern Research Trends*, 102
Harvard University, 29, 31–35, 62–63, 67, 111, 136n14
Hausfater, Glen, 113, 115
Hinde, Robert A., 104
Howard, Henry Elliot, 60, 62
howler monkeys, studies of, 8, 24, 51, 53, 55, 56–59, 62, 73, 111. *See also* Barro Colorado, Panama; Carpenter, Clarence Ray
homosexuality, 55, 76, 124; and female primates, 76; and male primates, 76
Huffington Post, 123
hunter-adventurer, 6, 23, 26, 27, 52, 99
hunter-collector, 8, 17, 27–28, 37, 39–40, 47. *See also* Burbridge, Ben
Huxley, Julian, 67

indigenous people: as assistants and researchers, 4, 6–9, 64, 105, 107–8, 120–22, 126–27, 129, 153n36; and colonial history of national parks, 109; as guides and porters, 6, 8, 65–67, 95, 97, 131n8; *indigenous* as term, 131n4, 131n8
infanticide, 96
Institute for Research in Tropical Behavior, 55
International Primatological Society, 3

Jackson Hole Wildlife Park, 84
Jane Goodall Institute Research Center (JGI), ix

Joe the Chimpanzee and Other Stories (anonymous), 21–22
Johns Hopkins University, 63
Johnson, Martin, 18, 64, 81
Johnson, Osa, 81; *Congorilla*, 18

Kajiado African District Council, 110
Kelly, Joan Morton, 79, 81
King Kong, 15, 18, 22, 24
Koch, Ludwig, 67
Kohler, Robert, 7, 72
Köhler, Wolfgang, 59, 61–62; *The Mentality of Apes*, 59
Kohts, Nadia, 36
Koko, 101
Kummer, Hans, 113

laboratory: and moral economy, 7; laboratory-like spaces, 56, 88, 127; limitations of, 46, 49, 51, 60, 72, 90; and standards for good science, 40, 53, 55, 71–72, 87; studies in the, 54–55, 67, 69, 86, 99, 102–4; and Robert Mearns Yerkes, 28–34, 42, 46. *See also* Barro Colorado; Carpenter, Clarence Ray; Harlow, Harry; Yale Laboratories of Primate Biology
Lancet, 16
Land and Live in the Jungle (air force training film), 65
Laura Spelman Rockefeller Memorial, 45
Leakey, Louis, 7
life history, 20, 92, 95, 113–14, 128–29. *See also* fitness
Life, 76–77, 124
linguistics, 6
London Zoological Gardens, 73
Lorenz, F. W., 85, 90
Lorenz, Konrad, 99; *On Aggression*, 99

Maasai, 1, 107, 109–10, 115, 119–21, 150n8. *See also* Amboseli National Park
Marinka, Gideon, 119
Markle-Columbia Primate Expedition, 75
Mary and John B. Markle Foundation, 75
Mason, William, 87–88
masturbation, 30, 55, 101, 124

mathematics, 39, 92, 104, 109, 111–13
McDougall, William, 53
Melincourt (Peacock), 12
Menzel, Emil W., 88
Merriam, John C., 47
MGM, 83. *See also* motion pictures
midwifery, 7
migration, 96, 108, 113
Mogambo (film), 83, 90
Morgan, Lloyd C., 20–22
motion pictures, 15, 37, 39, 98; for recording animal behavior, 40, 54, 58, 67, 71
Muruthi, Philip, 117, 122
Musée de L'Homme, Paris, 16
Museum of Comparative Zoology (Harvard University), 62
Mututua, Raphael, xi, xii, 1–2, 115–20, 121–22
myth, 4, 9, 11–12, 16, 27, 52, 123–24

National Geographic, 96, 98, 101, 123, 127, 149n31
National Geographic Society, 55
National Institutes of Health, 102, 111, 125
National Research Council, 30, 34, 54
National Science Foundation, 93
natural behavior: meaning of, 51, 58, 70–74, 78, 82–91; and "naturalistic," 34, 46, 49, 50, 87–89; and value of field studies, 3, 46–47, 60, 69, 95, 99
Naturalist Repository, The (Donovan), 22
naturalist tradition, 50, 58, 85, 87
Netherlands Indian Society for Nature Preservation, 62
New York Times, 35, 63, 98, 101, 123, 124
New York Zoological Society, 24, 84, 87, 93
Nissen, Henry, 49–52, 57
Nova, 123

observation: of captive primates, 5, 21, 23, 34, 36–42, 44, 54, 73, 82; challenges of field, 23, 48, 50–51, 78, 93; and credibility, 6, 25–27, 35, 38, 53, 66, 70–71, 85, 95, 103–5; long-term field, 4, 6, 7, 9, 28–29, 51, 56, 64, 91–92, 96, 108, 114–15, 122, 129; practices, 8, 53–54, 56–58, 66–67, 70, 78, 93, 104, 107, 109, 112–17, 127, 147n47, 151n14. *See also* adventure

narrative; dyadic method; habituation; indigenous; playback; sampling; women
One Small Step (film), 125
On the Fabric of the Human Body (Vesalius), 12
orangutans, 12, 28; studies of, 22–23, 60, 62, 64. *See also* Carpenter, Clarence Ray; Wallace, Alfred Russel
Orcutt, Edalee, 81
O'Reilly, John, 94
ornithology, and primate studies, 30, 56, 58, 61. *See also* birds
Osborn, Fairfield Henry, 47

Pasteur Institute of Kindia, 49
Pastrana, Julia, 16
Patterson, Francine, 101
Peabody Museum of Natural History, 20
Pennsylvania State University, 79
playback, 67–70
Playboy, 124
Poe, Edgar Allan, 12
Primate (film), 101. *See also* Wiseman, Frederick
primate folklore, 4–5, 16, 23–25, 27–28, 51, 72, 92, 98–99, 123
Primate Societies, 121
Princeton University, 108, 117, 122
professionalization, 3–5, 7, 51, 131n6
Project Nim, 125
pop primatology, 4, 92, 98, 101–2
provisioning, 103
Pusey, Anne, ix

Rausch, Harold L., 87
Regents Park, London, 21
rhesus macaques, 78, 87, 111, 124, 125; biomedical demand for, 74–75. *See also* Carpenter, Clarence Ray; Cayo Santiago, Puerto Rico; Harlow, Harry
Ringling Bros and Barnum and Bailey Circus, 18–19, 45. *See also* Congo (gorilla); Gargantua (Buddy)
Rockefeller Foundation, 8, 33, 45
Rodman, Peter, x
Rubenstein, Dan, 122
Rwanzagire, Reuben, 95

School of Tropical Medicine, San Juan, 75
sampling, 69, 112, 115; and Jeanne Altmann, 112, 127. *See also* observation
San Diego Zoo, 79; and Belle Benchley, 79; and gorilla studies, 81–83
Sapolsky, Robert, 114. *See also* darting
Sayialel, Serah, xi, xii, 1–2, 115–17, 119–22. *See also* Amboseli Baboon Project
Schaller, George, 24, 90, 92–97, 102, 108; *The Mountain Gorilla: Ecology and Behavior*, 94–95; *The Year of the Gorilla*, 94
Schaller, Kay, 93, 95–96
Scheirla, T. C., 85
Schrier, Allan M., 102–3
Schultz, Adolph H., 63–64
Science, 104, 113, 123
Scientific Monthly, 68
Scott, J.P., 84
Scott, Ken, 84
sex drive, 30, 54–55
sexology, 6
Silverback Gorilla Syndrome: Transforming Primitive Man (Hood), 125
Singer, Peter, 125
Smith College, 30
Smith, Philip, 75
Smithsonian Institution, 55, 83, 125
Social Science Research Council, 78, 81
Society for the Prevention of Cruelty to Animals, 74
sociology, 6, 63
Sparks, Richard, 32, 49, 92. *See also* animal welfare
Sports Illustrated, 94
Stollnitz, Fred, 102–3
Stone, Calvin P., 53–53
St. Paul Pioneer Press, 94
Strasser, Bruno, 7
Strier, Karen, 128
Sumner, Francis, 72

Tarzan: book series (Burroughs), 22; *Greystoke* film, 100
Temple Bar, 25
Terrace, Herb, 100
territoriality, 56, 61, 66

This American Life, 125
Time, 124
Tinbergen, Niko, 86
Tinklepaugh, Otto L., 51
Tomilin, M. I., 75
travel narratives, 51. *See also* adventure narratives
Tyson, Edward, 12–13, 24, 27

University of Alberta, 111
University of California, 111
University of Chicago, 86, 111
University of Kansas, 87
University of Michigan, 87
University of Nairobi, 122
University of Nevada, 100
University of Wisconsin, 84, 92–94, 102–3, 128
University of Wisconsin Primate Laboratory, 103
U.S. Fisheries and Wildlife Service, 125
U.S. National Museum (Siam), 63

Vanity Fair, 98
vocalization: of gibbons, 53; of gorillas, 24, 38, 92, 94, 96; recordings of, 65, 67–68, 82; and social function, 59–62, 69–70. *See also* communication
Vogue, 98

Wallace, Alfred Russel, 22–23, 56; *The Malay Archipelago*, 22
Ward's Natural Science Bulletin, 16
Washburn, Sherwood, 63–64
Washoe, 100–101
Warutere, Kinyua, xi, xii, 1–2, 115, 117, 120, 122. *See also* Amboseli Baboon Project
What Do Women Want? Adventures in the Science of Female Desire (Bergner), 124
Wheeler, William Morton, 35
Wild Kingdom, 98
Wildlife the World Over (Boulenger), 24
Willems, Edwin P., 87–88
Wilson, E. O., 111

Wiseman, Frederick, 101
White, Charles, 12
White, Richard Grant, 15
Whiting, Beatrice, 111
women: contributions to primate studies by, 6–7, 31, 35, 98, 126–28; and depictions in popular culture, 100–102, 126; and the human-animal boundary, 12, 16, 18; number in primatology, 7–9, 149n34. *See also* Abreu, Rosalia; Alberts, Susan; Altmann, Jeanne; Baartman, Sara; Benchley, Belle Jennings; Fossey, Dian; Goodall, Jane; Pastrana, Julia; Sayialel, Serah

Yale Laboratories of Primate Biology, 30–31, 46–47, 49–50, 103, 135n6, 141n19, 144n5. *See also* Bingham, Harold; Carpenter, Clarence Ray; Nissen, Henry; Yerkes, Ada; Yerkes, Robert Mearns
Yerkes, Ada, 28–31, 33
Yerkes, David, 31
Yerkes, Robert Mearns: *Almost Human*, 35–36; and captive studies, 28–30, 32; and Chim and Panzee, 34–35; and collaboration with hunter-adventurer Ben Burbridge, 28, 37–40; and collaborations with women, 8, 28, 30–31, 35–37, 79–82; and communication studies, 59–62; and funding for primate studies, 28, 30, 33, 124; on the "Gorilla Trail," 78–79, 81, 86, 93; and interdisciplinarity, 5, 29; *An Introduction to Psychology*, 29; outside the laboratory, 33–40, 45–46; postdoctoral fellows, 6, 29; relationship with primate subjects, 32, 41–42; and value of field studies, 28, 71, 73, 90–91, 138n67; views of naturalness, 78, 82–84, 86, 90–91, 141n19. *See also* Yale Laboratories of Primate Biology
Yerkes Regional Primate Research Center, 109

zoos. *See* Bronx Zoo; London Zoological Gardens; San Diego Zoo
Zuckerman, Solly, 73, 144n5